Biological Processes and Techniques in Zero Waste Management

Biological Processes and Techniques in Zero Waste Management

Garima Awasthi

Amity Institute of Biotechnology, Amity University Uttar Pradesh, Lucknow Campus, Lucknow, U.P., 226028, India

Mohit Nigam

Department of Biotechnology, School of Pharmaceutical and Biological Sciences, Harcourt Butler Technical University, Kanpur, U.P., 208002, India

WILEY VCH

Authors

Garima Awasthi
Amity University Uttar Pradesh, Amity Institute of Biotechnology, Lucknow Campus, Lucknow 226 028 UP, India

Mohit Nigam
Department of Biotechnology,
School of Pharmaceutical and Biological Sciences,
Harcourt Butler Technical University,
Kanpur, U.P., 208002, India

Cover Design: Wiley
Cover Image: © Elesey/Shutterstock

Typesetting: Lumina Datamatics, Madurai, India
Printing and Binding: CPI Group (UK) Ltd, Croydon, CR0 4YY

Library of Congress Card No.: applied for

British Library Cataloguing-in-Publication Data
A catalogue record for this book is available from the British Library.

Bibliographic information published bythe Deutsche Nationalbibliothek
The Deutsche Nationalbibliothek lists this publication in the Deutsche Nationalbibliografie; detailed bibliographic data are available on the Internet at <http://dnb.d-nb.de>.

Print ISBN: 9783527355198
ePDF ISBN: 9783527852321
ePub ISBN: 9783527852314
oBook ISBN: 9783527852338

C9783527852321_250226

Contents

Preface

The escalating global waste crisis is one of the defining environmental challenges of our time. Driven by industrial growth, rapid urbanization, and unsustainable consumption patterns, the volume and complexity of waste streams are expanding at an alarming pace. Traditional disposal practices, landfilling, open dumping, and incineration, are proving inadequate, contributing to greenhouse gas emissions, resource depletion, and ecological degradation. In response, the zero-waste philosophy has emerged as a transformative approach, aiming not only to minimize waste generation but to recover, recycle, and repurpose materials in ways that create environmental, economic, and social value. This book brings together scientific insights, technological innovations, and practical applications that illustrate how waste can be reimagined as a valuable resource. The chapters traverse a wide spectrum, from microbial valorization of organic waste, bio-based industrial raw materials, and innovative recycling techniques to policy frameworks, community-level initiatives, and industrial symbiosis models. By integrating perspectives from microbiology, process engineering, materials science, and circular economy strategies, the work offers a comprehensive roadmap for transitioning from linear "take-make-dispose" systems to regenerative, closed-loop models. Each section emphasizes actionable solutions. Readers will discover how bioconversion, thermochemical processing, chemical recycling, and advanced separation technologies can convert waste into biofuels, bioplastics, fertilizers, renewable energy, and high-value chemicals. Case studies from around the world demonstrate the scalability of these approaches, whether in decentralized composting in urban neighborhoods, large-scale biogas plants in agricultural hubs, or precision fermentation start-ups turning food waste into sustainable proteins. The intended audience is as diverse as the subject itself: researchers, engineers, policymakers, entrepreneurs, educators, and community leaders. Whether the goal is to advance scientific understanding, design policy interventions, or implement practical waste-to-resource projects, the insights compiled here aim to inspire informed action. The journey toward zero waste demands systemic change, rethinking product design, consumption patterns, supply chains, and waste management infrastructures. It also requires collaborative effort across disciplines, sectors, and borders. By combining innovation with responsibility, we can shift from viewing waste as an inevitable burden to recognizing it as a cornerstone of a resilient, sustainable future. This book is an invitation to be part of that transformation.

Preface

The escalating global waste crisis is one of the defining environmental challenges of our time. Driven by industrial growth, rapid urbanization, and unsustainable consumption patterns, the volume and complexity of waste streams are expanding at an alarming pace. Landfills and disposal practices (landfilling, open dumping, and incineration) are proving inadequate, contributing to greenhouse gas emissions, resource depletion, and ecological degradation. In contrast, the emerging philosophy [illegible] environmental and social value. This book brings together scientific insights, technological innovations, and practical applications that illustrate how waste can be reimagined as a valuable resource. The chapters traverse a wide spectrum, from microbial valorization of organic waste [illegible] industrial raw materials, and [illegible] plastics, pyrolysis, fertilizers, renewable energy, and high-value chemicals. Case studies from around the world demonstrate the scalability of these approaches, whether in decentralized composting in urban neighborhoods, large-scale biogas plants in agricultural hubs, or precision fermentation start-ups turning food waste into sustainable proteins. The intended audience is as diverse as the subject itself: researchers, engineers, policymakers, entrepreneurs, educators, and community leaders. Whether the goal is to advance scientific understanding, design policy interventions, or implement practical waste-to-resource projects, the insights compiled here aim to inspire informed action. The journey toward zero waste demands systemic change: rethinking product design, consumption patterns, supply chains, and waste management infrastructures. It also requires collaborative effort across disciplines, sectors, and borders. By combining innovation with responsibility, we can shift from viewing waste as an inevitable burden to recognizing it as a cornerstone of a resilient, sustainable future. This book is an invitation to be part of that transformation.

Part I

Basic Biological Sources, Tools, and Techniques

1

Microbes: Basic Tools and Techniques for Valorization of Waste

1.1 Introduction

The global waste crisis has escalated into a pressing environmental, economic, and social issue. With increasing industrialization, urbanization, and consumerism, the volume and complexity of waste generated globally has reached unprecedented levels. According to the World Bank, over 2.01 billion tonnes of municipal solid waste (MSW) are produced annually, and this is expected to grow to 3.4 billion tonnes by 2050. Traditional methods of waste disposal, such as landfilling and incineration, have proven inadequate due to their environmental ramifications, including greenhouse gas emissions, leachate generation, and long-term land usage. Consequently, the transition from waste management to waste valorization, converting waste into useful resources, has become an imperative.

In this transformative approach, microorganisms have emerged as vital agents of change. Their metabolic versatility, adaptability to extreme environments, and ability to transform, degrade, or synthesize a wide range of organic and inorganic compounds make them indispensable in sustainable waste valorization technologies. Microbial systems have long been utilized in traditional fermentation and composting practices, but modern advances in microbiology, molecular biology, and systems biotechnology have significantly broadened their application.

This study provides an in-depth examination of how microbes serve as essential tools for the biological valorization of waste. Emphasis is placed on microbial diversity, metabolic pathways, and cutting-edge techniques including microbial consortia, synthetic biology, enzymatic biotransformation, and bioreactor designs. The integration of these approaches forms the basis for efficient, cost-effective, and environmentally friendly valorization systems. The following sections will explore the necessity of these methods, the microbiological underpinnings of waste valorization, and the state-of-the-art microbial techniques currently driving innovation in this field.

Biological Processes and Techniques in Zero Waste Management, First Edition. Garima Awasthi and Mohit Nigam.

By understanding the potential of microbes not merely as waste degraders but as resource generators, we lay the groundwork for a circular bioeconomy that aligns with global sustainability goals. Microbial valorization technologies include a broad spectrum of techniques such as aerobic composting, anaerobic digestion (AD), solid-state and submerged fermentation (SmF), microbial fuel cells (MFCs), enzymatic transformation, and phototrophic systems using algae and cyanobacteria. Each of these processes utilizes specific microbial species or consortia capable of transforming particular waste types. For instance, composting relies on aerobic bacteria and fungi for the degradation of organic waste into stable humus, while AD employs methanogens and syntrophic bacteria to produce biogas from high-moisture substrates. Fermentation techniques, both solid and submerged, are employed to manufacture enzymes, acids, alcohols, and even biodegradable plastics from agricultural or food processing waste.

Moreover, the rise of synthetic biology, CRISPR-based gene editing, and metagenomic approaches has further expanded the potential of microbial waste valorization. These tools have enabled the development of engineered microbial strains with enhanced substrate specificity, higher product yields, and greater tolerance to toxic environments. For example, engineered *Escherichia coli, Corynebacterium glutamicum*, and yeast strains are now used to convert food and plastic waste into value-added chemicals and proteins. Synthetic microbial consortia, designed with tailored functional capabilities, offer improved control over complex biodegradation and biosynthesis pathways. Beyond the laboratory, the deployment of microbial technologies at the industrial and community levels has already begun to reshape the landscape of waste management. From decentralized composting units in urban residential areas to industrial-scale biogas plants and precision fermentation startups, microbial processes are proving to be scalable, cost-effective, and environmentally responsible solutions.

In this chapter, we explore the fundamental tools and techniques that leverage microbial systems for waste valorization. By understanding the mechanisms, applications, and innovations in microbial technologies, we can uncover the immense potential of these microscopic agents in addressing one of the most pressing global challenges: sustainable waste management. This knowledge not only contributes to environmental preservation but also supports the development of a circular economy wherein waste is no longer a burden but a valuable resource.

1.2 Microbial Diversity and Metabolic Potential

Microorganisms represent an extraordinary reservoir of biochemical and ecological potential. Their diversity spans across bacteria, archaea, fungi, actinomycetes, and microalgae, each group harboring unique capabilities that make them instrumental in the degradation, transformation, and conversion of various waste types. A profound understanding of microbial diversity is the foundational requirement for tailoring biotechnological processes aimed at waste valorization.

1.2.1 Bacteria

Bacteria are the most studied and widely utilized microorganisms in waste valorization. Genera such as *Pseudomonas, Bacillus, Clostridium*, and *Acinetobacter* are known for their enzymatic richness and metabolic flexibility. These microbes can degrade a vast array of pollutants including hydrocarbons, plastics, pesticides, heavy metals, and complex organic matter.

Pseudomonas putida, for instance, exhibits metabolic plasticity that allows it to utilize aromatic compounds and xenobiotics, making it highly suitable for the biotransformation of industrial and municipal waste. *Clostridium* spp. are obligate anaerobes that ferment organic substrates into valuable products like hydrogen, acetate, and butanol under strictly anoxic conditions.

1.2.2 Fungi

Fungi, especially filamentous species like *Aspergillus, Trichoderma*, and *Penicillium*, play a vital role in the breakdown of lignocellulosic waste due to their robust extracellular enzyme systems. These include cellulases, xylanases, laccases, and peroxidases. Fungi are particularly effective in composting systems and in the treatment of agro-industrial residues rich in cellulose and hemicellulose.

White-rot fungi such as *Phanerochaete chrysosporium* are capable of mineralizing lignin, a complex and recalcitrant biopolymer, making them indispensable in the pretreatment of plant biomass for biofuel production.

1.2.3 Actinomycetes

Actinomycetes, a group of filamentous, Gram-positive bacteria, bridge the morphological characteristics of fungi and bacteria. *Streptomyces, Nocardia*, and *Micromonospora* are examples of actinomycetes that produce a wide range of secondary metabolites and extracellular enzymes. Their role in degrading complex organic materials in soil and compost environments makes them valuable in solid waste valorization systems.

1.2.4 Archaea

Although less explored, archaea, especially methanogenic archaea, play a crucial role in AD. These microbes are responsible for the final step in the breakdown of organic matter, converting acetate, hydrogen, and CO_2 into methane. *Methanosarcina* and *Methanobacterium* are examples of genera involved in biogas production.

Extremophilic archaea also offer potential in treating high-salinity or high-temperature waste streams, expanding the operational boundaries of biological systems.

1.2.5 Microalgae

Microalgae such as *Chlorella, Scenedesmus*, and *Spirulina* have shown remarkable capabilities in wastewater treatment and CO_2 sequestration. They assimilate nutrients such as nitrogen and phosphorus from waste streams and simultaneously produce biomass that can be converted into biofuels, feed, or nutraceuticals.

The photosynthetic nature of microalgae makes them ideal for integration into photobioreactor systems where light-driven valorization processes are desired.

1.2.6 Mixed Microbial Consortia

In natural and engineered systems, waste valorization is rarely the job of a single organism. Instead, microbial consortia, complex communities comprising various species, exhibit synergistic interactions that enhance degradation efficiency, broaden substrate specificity, and increase process resilience. These consortia can be harnessed from natural environments or synthetically assembled for targeted bioprocesses.

Metagenomic analyses have revolutionized our ability to explore and exploit these consortia, enabling the discovery of novel metabolic pathways and functional genes involved in waste bioconversion.

1.3 Need for Microbial Valorization Techniques

The demand for efficient, sustainable, and eco-friendly waste treatment technologies has never been greater. As global waste generation continues to rise, conventional disposal methods—landfilling, incineration, and chemical treatment—pose significant limitations and environmental risks. Landfills are associated with long-term contamination of soil and groundwater, while incineration releases hazardous air pollutants and contributes to climate change through greenhouse gas emissions. Chemical treatments, though sometimes effective, are often expensive, energy-intensive, and may result in secondary pollution. These limitations underscore the need for alternative waste valorization strategies, with microbial methods offering a particularly promising solution.

1.3.1 Environmental Necessity

Microbial valorization techniques provide a natural and sustainable method for breaking down and transforming waste materials into nontoxic or even valuable compounds. The biological nature of these processes allows for minimal energy input and lower environmental impact compared to thermal or chemical alternatives. Microbes are capable of acting on complex and heterogeneous waste streams, often under mild operational conditions, thereby reducing both ecological and economic burdens [1].

In wastewater treatment, for example, microbial processes are used not only for the removal of organic pollutants and nutrients but also for the production of biogas and biofertilizers, creating a closed-loop system. Similarly, composting and AD of MSW convert organic material into useful end products, such as compost and methane, respectively, which have direct applications in agriculture and energy generation [2].

1.3.2 Economic Viability

Economically, microbial valorization is increasingly viable as advances in biotechnology improve microbial yields, enzyme efficiencies, and reactor designs. Engineered strains can be developed to produce high-value compounds such as biosurfactants, organic acids, and bioplastics from low-cost substrates derived from waste. The ability to generate value-added products from what was once considered waste transforms cost centers into potential revenue streams [3].

Moreover, decentralized microbial waste valorization units can be integrated into rural and urban waste management infrastructures. These systems offer scalable and flexible solutions, especially in regions lacking centralized waste treatment facilities. Community-level composting or biogas systems powered by local microbial consortia exemplify the adaptability and affordability of microbial technologies [4].

1.3.3 Regulatory and Policy Drivers

National and international policies increasingly favor sustainable waste treatment technologies. The European Union's Circular Economy Action Plan, the United Nations Sustainable Development Goals (SDGs), and country-specific regulations on landfill reduction and renewable energy all support the development and adoption of microbial valorization technologies. In particular, SDG 12 (Responsible Consumption and Production) and SDG 13 (Climate Action) align closely with the goals of microbial waste valorization [5].

1.3.4 Resilience and Adaptability

One of the greatest advantages of microbial systems is their adaptability. Microorganisms can evolve or be engineered to handle diverse waste compositions, tolerate harsh environmental conditions, and resist inhibitors commonly found in industrial effluents. Their resilience makes them suitable for dynamic waste environments where chemical or physical methods may fail or be uneconomical.

Furthermore, microbial communities can be fine-tuned through selective enrichment, co-culture development, and systems biology approaches to optimize degradation pathways and product yields. This adaptability is key in meeting the complex and evolving challenges posed by modern waste streams, which often contain novel contaminants such as pharmaceuticals, microplastics, and heavy metals [6].

1.3.5 Circular Bioeconomy and Sustainability Goals

The concept of the circular bioeconomy promotes the reuse, recycling, and biological upgrading of waste streams into resources. Microbial valorization sits at the heart of this concept, offering technologies that not only mitigate environmental harm but also generate renewable materials and energy. From producing biohydrogen from food waste to synthesizing biopolymers from agricultural residues, microbial systems can significantly contribute to reducing dependency on fossil resources and enhancing material circularity [7]. In summary, the pressing environmental issues, economic opportunities, regulatory mandates, and sustainability goals collectively establish a strong need for microbial valorization technologies. In the following sections, we will delve into the specific microbial techniques and bioprocesses that underpin this transformative approach to waste management.

1.4 Microbial Techniques for Waste Valorization

Microorganisms offer a powerful and sustainable approach to waste valorization due to their vast metabolic diversity and ability to convert organic and inorganic waste into valuable products. These techniques span from well-established methods like composting and fermentation to more recent innovations such as MFCs and synthetic biology-based processes. Each method exploits the unique physiological characteristics of microbes to address various waste types while contributing to environmental sustainability and resource recovery. The following is a comprehensive overview of the key microbial techniques employed in waste valorization.

1.4.1 Aerobic Composting

Aerobic composting is one of the oldest and most widely practiced methods for converting organic waste into a nutrient-rich soil amendment. The process involves the decomposition of organic matter in the presence of oxygen, facilitated by aerobic microorganisms such as bacteria and fungi.

During composting, microbial activity generates heat, which supports different temperature-dependent microbial communities. Initially, mesophilic bacteria dominate, followed by thermophilic organisms that degrade complex compounds like cellulose and lignin. As the process matures, mesophilic microbes reappear to stabilize the compost.

This technique is commonly used for household food waste, agricultural residues, and garden waste. Compost improves soil health by increasing nutrient content, water retention, and microbial diversity. Recent improvements include the use of microbial inoculants, optimized aeration, and the addition of bulking agents like biochar to enhance efficiency and reduce pathogen load [8].

1.4.2 Anaerobic Digestion

AD is a microbial process where organic materials are broken down in the absence of oxygen, producing biogas—a mixture mainly consisting of methane and carbon dioxide and a nutrient-rich digestate.

The process occurs in four main steps: hydrolysis (breaking down complex polymers), acidogenesis (formation of volatile fatty acids), acetogenesis (conversion of acids to acetate and hydrogen), and methanogenesis (methane production by archaea). This complex microbial interaction is carried out by a consortium of bacteria and methanogens.

AD is suitable for treating high-moisture organic wastes such as food scraps, animal manure, sewage sludge, and industrial effluents. Technological advancements like co-digestion strategies, microbial bioaugmentation, and innovations in reactor designs (e.g., Upflow Anaerobic Sludge Blanket, Anaerobic Membrane Bioreactors) have significantly improved biogas yield and operational stability [9].

1.4.3 Solid-state Fermentation

Solid-state fermentation (SSF) is a microbial cultivation technique conducted on solid materials with minimal or no free-flowing water. It is particularly useful for converting agro-industrial residues and lignocellulosic biomass into valuable bio-products.

Fungi, especially filamentous species like *Aspergillus* and *Trichoderma*, are commonly used in SSF due to their ability to penetrate solid substrates and secrete extracellular enzymes. SSF is employed for producing enzymes (such as cellulase and amylase), organic acids, ethanol, and feed supplements.

Recent enhancements in SSF include substrate pretreatment to increase microbial accessibility, the use of genetically improved strains, and the adoption of bioreactor configurations like tray and packed bed systems for industrial-scale applications.

1.4.4 Submerged Fermentation

SmF is a traditional technique where microorganisms are cultivated in a liquid nutrient medium under controlled environmental conditions. It is widely used in industrial microbiology for the production of metabolites and bio-based chemicals.

Common microbes include *Saccharomyces cerevisiae, Lactobacillus*, and *Zymomonas* species. SmF is particularly suitable for converting liquid and soluble waste streams into ethanol, citric acid, lactic acid, antibiotics, and bioplastics.

Key improvements in SmF processes include fed-batch and continuous operation modes, advanced pH and oxygen control systems, and improved mixing and aeration designs. These developments have helped scale up the process efficiently and economically [10].

1.4.5 Microbial Fuel Cells

MFCs are bioelectrochemical systems that utilize the metabolic activity of electroactive bacteria to convert organic waste into electrical energy. In an MFC, microbes oxidize waste materials in the anode chamber, releasing electrons that travel through an external circuit to the cathode, thus generating electricity.

Electroactive bacteria such as *Geobacter* and *Shewanella* play a pivotal role in this process. MFCs offer dual benefits: waste treatment and energy recovery, making them an attractive solution for treating wastewater and low-strength organic waste streams.

Ongoing research is focused on improving electrode materials, enhancing electron transfer efficiency, and optimizing reactor architecture to make MFCs more scalable and economically feasible [11].

1.4.6 Enzymatic Biotransformation

Enzymatic biotransformation involves the use of microbial enzymes to catalyze the breakdown or conversion of specific components within waste streams into simpler or value-added products. This process is highly specific, efficient, and environmentally benign.

Commonly used enzymes include cellulases (for lignocellulose degradation), lipases (for oil waste), proteases (for protein-rich waste), and amylases (for starch-containing materials). These enzymes can be used in free or immobilized forms for continuous operations.

Applications span industries including textile (dye degradation), food (liquefaction of waste), biofuels (biomass hydrolysis), and detergents. The integration of metagenomic approaches, enzyme immobilization techniques, and protein engineering has further improved enzyme efficiency, stability, and specificity [12].

1.4.7 Synthetic Microbial Consortia

Synthetic microbial consortia involve the deliberate assembly of multiple microbial strains with complementary metabolic functions to enhance the degradation or conversion of complex waste streams. Each strain is engineered or selected to perform specific roles, such as hydrolysis, detoxification, or fermentation.

Such consortia are particularly useful for processing heterogeneous or toxic waste materials like plastics, e-waste, and industrial effluents. The coordinated activity of diverse microbes ensures a more robust and flexible process compared to single-strain systems.

Challenges include maintaining stability and avoiding competition among members, but advances in synthetic biology and microbial ecology are enabling better control over community composition and function [13].

1.4.8 Precision Fermentation

Precision fermentation leverages genetic engineering and synthetic biology to program microorganisms for the production of high-value compounds using waste as the primary feedstock. Using tools such as CRISPR, microbes can be engineered to produce proteins, enzymes, bioplastics, and nutraceuticals.

Engineered strains of *E. coli*, *C. glutamicum*, and various yeasts are commonly used in this technique. Precision fermentation allows for tailored biosynthesis, enabling the transformation of food waste, agricultural by-products, or industrial effluents into targeted products.

This approach is gaining traction in the biotechnology industry, with companies developing microbial platforms to convert organic waste into functional foods, single-cell proteins, and specialty chemicals [14].

1.4.9 Phototrophic Waste Valorization

Phototrophic waste valorization utilizes photosynthetic microorganisms such as microalgae and cyanobacteria to convert light energy and waste-derived nutrients into biomass and valuable biochemicals. These organisms assimilate nitrogen, phosphorus, and carbon dioxide, making them ideal for wastewater treatment.

Microalgae cultivation systems such as open raceway ponds or closed photobioreactors can produce biofuels (lipids), pigments, antioxidants, and protein-rich biomass. Coupling these systems with wastewater treatment or carbon capture technologies offers a sustainable, dual-purpose approach to waste valorization and environmental protection.

1.5 Case Studies and Real-world Applications

Microbial waste valorization has rapidly moved beyond laboratory research to become a cornerstone of sustainable waste management strategies across industries and geographies. The following case studies demonstrate the practical implementation of microbial technologies, highlighting their scalability, socio-economic benefits, and environmental impact. Each case reflects how tailored microbial systems can transform diverse waste streams into valuable resources [15].

1.5.1 MSW Management in Pune, India

The Pune Municipal Corporation (PMC), in partnership with a local biotechnology startup, launched a decentralized composting initiative to tackle the city's growing organic waste problem. The system primarily targets biodegradable fractions of MSW, including vegetable peels, kitchen scraps, and market refuse.

- **Technology Deployed:** Enclosed, in-vessel aerobic composting units with automated temperature and moisture monitoring systems ensured optimal microbial activity and minimized odor.
- **Results:** Over 800 tons of organic wastes are processed monthly, yielding nutrient-rich compost that is sold to local farmers and gardeners. The project significantly reduced open dumping, controlled vector proliferation, and boosted citizen engagement in waste segregation and sustainability.

This decentralized model has been recognized as a replicable solution for other urban centers struggling with organic waste accumulation [16].

1.5.2 Biogas Production from Dairy Waste in Denmark

In the dairy-intensive region of Jutland, Denmark, a state-of-the-art AD facility was established to convert high-moisture dairy by-products such as whey, manure, and sludge into renewable energy.

- **Microbial Consortia:** Methanogenic archaea like *Methanosarcina barkeri* and acidogenic bacteria were strategically employed for multistage digestion hydrolysis, acidogenesis, acetogenesis, and methanogenesis.
- **Co-digestion Strategy:** Glycerol from biodiesel plants and brewery waste were added to the mix to enhance carbon availability and increase biogas output.
- **Performance:** The plant generates more than 20 MW of electricity daily, which is fed into the national grid. The residual digestate is rich in nutrients and is used as a liquid fertilizer, effectively replacing synthetic agrochemicals and reducing greenhouse gas emissions.

This facility exemplifies circular bioeconomy in action, integrating waste management with energy and agriculture [17].

1.5.3 Agro-waste Biorefinery in Brazil

A sugar mill in São Paulo, Brazil, has developed an integrated biorefinery to valorize agro-industrial residues, specifically sugarcane bagasse and vinasse a by-product of ethanol distillation.

- **Microbial Agents:** *S. cerevisiae* was used for ethanol fermentation, *Aspergillus niger* for cellulase production via SSF, and *Lactobacillus plantarum* for organic acid synthesis through SmF.
- **Innovative Infrastructure:** Modular bioreactors allowed for simultaneous, parallel fermentations, optimizing energy use and resource allocation.
- **Impact:** The biorefinery achieved a 30% reduction in effluent discharge, decreased environmental compliance costs, and opened new revenue streams through the sale of industrial enzymes and organic acids.

This model showcases how traditional agricultural industries can diversify into value-added bioproducts using microbial bioprocessing [18].

1.5.4 Microbial Plastic Biodegradation Pilot in Japan

A pioneering pilot program in Osaka, Japan, is exploring microbial solutions for plastic waste, focusing on the biodegradation of polyethylene terephthalate (PET) and polyurethane (PU).

- **Biological Systems:** The key players are *Ideonella sakaiensis*, known for its PETase enzyme activity, and genetically modified strains of *P. putida*, capable of degrading PU.
- **Operational Strategy:** Controlled bioreactor systems operating at moderate temperatures, along with enzyme supplementation, were used to maintain degradation efficiency.

- **Achievements:** PET degradation reached 60% within 6 weeks—a significant improvement compared to passive landfill decomposition. Further scale-up studies are underway, including reactor design improvements and enzyme engineering for increased thermostability and substrate specificity.

This initiative sets the foundation for future industrial-scale microbial plastic recycling facilities [19].

1.5.5 MFCs in Slum Communities of Kenya

In Nairobi, Kenya, an NGO-led program deployed low-cost MFCs to address two major challenges: poor sanitation and limited electricity access in informal settlements.

- **Design Features:** Dual-chamber MFCs were built using inexpensive, locally available materials. Organic waste inputs included cow dung, kitchen waste, and sewage sludge.
- **Microbial Community:** Exoelectrogenic microbes such as *Shewanella oneidensis* and enriched local bacterial consortia facilitated the conversion of organic matter into electrical energy.
- **Outcomes:** The system produced sufficient electricity to power basic appliances like LED lamps and mobile phone chargers while simultaneously reducing organic waste and pathogens in the environment. It also created opportunities for local youth in the maintenance and operation of the units [20].

1.5.6 Photobioreactor Installations in UAE Greenhouses

As part of the United Arab Emirates' (UAE) effort to integrate sustainable agriculture with environmental stewardship, microalgae-based photobioreactors have been installed in commercial greenhouses.

- **Microalgae Species:** *Chlorella vulgaris* and *Spirulina platensis* were cultivated using sunlight, nutrient-rich wastewater, and CO_2 from greenhouse exhaust systems.
- **System Configuration:** Closed-loop photobioreactors coupled with desalination brine and nutrient recycling streams allowed dual-purpose operation biomass generation and wastewater treatment.
- **Benefits:** Over 80% nutrient recovery efficiency was achieved, and the harvested algal biomass was used as animal feed supplements and biofertilizers. Additionally, the system contributed to water conservation and reduced reliance on synthetic inputs [21].

1.5.7 Precision Fermentation of Food Waste in the United States

In California, a biotechnology startup is revolutionizing food waste valorization using precision fermentation to produce high-value protein products.

- **Technological Approach:** CRISPR-edited strains of *C. glutamicum* were used in fed-batch bioreactors to convert food processing residues into functional proteins.
- **Feedstock Used:** Potato peels, fruit pulp, and cheese whey served as low-cost, nutrient-rich substrates.
- **Production Metrics:** The facility consistently produces 200 kg of mycoprotein-rich biomass daily. The product, having met FDA food safety standards, is being commercialized as a sustainable protein source for dietary and functional food applications.

This example highlights how synthetic biology and microbial fermentation can drive the transition toward zero-waste food systems and alternative protein markets. These real-world examples reflect the dynamic and diverse applications of microbial waste valorization. By leveraging specific microbial capabilities and appropriate biotechnological systems, industries and communities can transform waste into wealth while advancing environmental sustainability and economic resilience [22].

1.6 Challenges, Opportunities, and Future Directions

Despite the demonstrated potential and increasing application of microbial waste valorization techniques, several technical, economic, and regulatory challenges persist. Addressing these hurdles is essential for scaling up microbial technologies and integrating them into mainstream waste management and resource recovery frameworks. This section explores the key challenges, emerging opportunities, and future trajectories for microbial valorization systems [23].

1.6.1 Challenges in Microbial Waste Valorization

1.6.1.1 Heterogeneity and Complexity of Waste Streams

Waste materials often possess varying physical, chemical, and biological compositions, which influence microbial degradation efficiency. High levels of inhibitors, toxic compounds, or recalcitrant molecules in industrial or plastic waste can impair microbial growth and activity.

1.6.1.2 Process Optimization and Scale-up

Laboratory-scale results often fail to replicate at the pilot or industrial level due to limitations in mass transfer, oxygen diffusion, temperature control, and reactor design. Maintaining microbial stability and performance over long-term continuous operation also presents technical bottlenecks.

1.6.1.3 Economic Viability

Although microbial processes are environmentally sound, the costs associated with bioreactor construction, process monitoring, microbial strain development, and product purification can be substantial. Competing with fossil-derived products or well-established chemical methods remains an economic hurdle.

1.6.1.4 Regulatory and Biosafety Concerns

The use of genetically modified organisms (GMOs) in valorization processes poses biosafety concerns and is subject to stringent regulatory scrutiny in many countries. Waste-derived products also face challenges in obtaining certifications for food, pharma, or environmental use [24].

1.6.1.5 Public Perception and Market Acceptance

Consumer and stakeholder acceptance of microbial and waste-derived products may be hindered by misconceptions or safety concerns. Effective science communication and transparent life-cycle assessments are necessary to build trust.

1.6.2 Opportunities for Innovation and Expansion

1.6.2.1 Synthetic Biology and Metabolic Engineering

CRISPR and other synthetic biology tools enable precise control over microbial metabolic pathways, enhancing substrate utilization, tolerance to inhibitors, and production of targeted compounds. This is opening new frontiers in customized waste valorization.

1.6.2.2 AI-driven Process Optimization

Machine learning and artificial intelligence are being integrated into microbial process design, allowing predictive modeling of microbial kinetics, optimization of fermentation parameters, and early detection of process anomalies.

1.6.2.3 Integration with Renewable Energy Systems

Coupling microbial valorization units with solar, wind, or geothermal energy sources can improve sustainability and reduce dependency on fossil fuels. For instance, photobioreactors utilizing solar input and waste CO_2 streams exemplify synergistic circular systems.

1.6.2.4 Biorefineries and Circular Bioeconomy Models

Emerging concepts such as biorefineries utilize multiple microbial techniques in tandem to convert diverse waste streams into a spectrum of products energy, bio-based materials, and fine chemicals enabling maximum resource efficiency.

1.6.2.5 Policy Support and Funding Incentives

Growing global emphasis on sustainable development has led to favorable policy landscapes. Government grants, carbon credits, and international climate financing mechanisms can support R&D, pilot installations, and technology transfer [25].

1.6.3 Future Directions

The future of microbial waste valorization lies in the development of next-generation microbial platforms, particularly those involving engineered extremophiles and synthetic microbial consortia capable of tackling complex and hazardous waste streams such as nuclear residues and electronic waste. A deeper understanding of microbial systems will be achieved through the integration of omics technologies—genomics, transcriptomics, proteomics, and metabolomics—allowing for advanced characterization of microbial communities and enabling the design of more efficient and resilient bioprocesses. Furthermore, hybrid systems that combine microbial techniques with physical or chemical methods, such as bioelectrochemical systems or thermo-bio pretreatment strategies, offer promising synergies for enhancing waste degradation efficiency. As waste challenges vary across regions and industries, there is also a growing need for personalized microbial solutions tailored to specific community or sectoral needs. To support this global transition, collaborative international platforms focused on microbial technology exchange, capacity building, and the creation of open-source microbial strain repositories will be essential for accelerating innovation and ensuring equitable access to sustainable waste management technologies.

1.7 Conclusion

Microbial waste valorization stands at the nexus of environmental sustainability, technological innovation, and socio-economic development. This chapter has demonstrated that microbes, nature's most efficient decomposers and chemical transformers, are essential agents in converting a wide variety of waste streams into valuable bio-products, including compost, biofuels, enzymes, organic acids, and bioenergy.

Through an in-depth exploration of microbial techniques such as composting, AD, fermentation (solid-state and submerged), MFCs, synthetic consortia, and precision fermentation, it is evident that microbial strategies offer scalable, versatile, and ecologically sound solutions for waste valorization. These approaches not only mitigate the environmental hazards of untreated waste but also open new value chains in bioeconomy sectors. Case studies from across the globe ranging from decentralized composting in India to precision fermentation in the United States have further illustrated the practicality, adaptability, and economic potential of microbial valorization technologies. The emergence of advanced tools like synthetic biology, machine learning, and omics-based profiling is further revolutionizing the way microbial processes are designed, optimized, and applied.

Nevertheless, the deployment of microbial valorization at industrial scale is not without challenges. Waste heterogeneity, regulatory restrictions, economic constraints, and public perception remain pressing issues. Yet, the continuous advancements in microbial engineering, bioprocessing equipment, and supportive policy frameworks present unprecedented opportunities for global adoption. Looking ahead, interdisciplinary collaboration among scientists, engineers, policymakers, entrepreneurs, and civil society will be critical in integrating microbial technologies into mainstream waste management systems. Education and outreach efforts will also be essential to shift societal perceptions and build acceptance for waste-derived products.

In conclusion, microbes offer a compelling blueprint for the sustainable transformation of waste into wealth. Their integration into circular economy models holds the promise of addressing pressing environmental issues while creating green jobs, reducing resource dependency, and fostering resilient communities.

References

1 Wang, L., Zhang, D., Liu, C., et al. (2023). Microbial strategies for waste valorization: recent advances and challenges. *Biotechnol Adv.* 61: 108129.

2 Singh, A., Pathak, V., and Yadav, S. (2023). Advances in anaerobic digestion and composting for municipal solid waste valorization. *J Environ Manage.* 326: 116745.

3 Ramesh, A., Kumari, P., and Singh, R. (2022). Engineered microbial systems for biorefinery applications. *Trends Biotechnol.* 40 (12): 1283–1295.

4 Das, S., Bhunia, B., and Dey, A. (2023). Decentralized microbial bioprocesses for sustainable waste management. *Renew Sustain Energy Rev.* 173: 113131.

5 European Commission (2022). Circular economy action plan. https://ec.europa.eu (accessed 12 June 2025).

6 Patel, R., Manna, S., and Bhattacharya, A. (2023). Microbial remediation of emerging pollutants in waste streams. *Crit Rev Biotechnol.* 43 (1): 19–34.

7 Jha, S., Sharma, V., and Rawat, R. (2022). Circular bioeconomy using microbial valorization of organic waste. *Bioresour Technol.* 360: 127619.

8 Sharma, V.K., Dubey, S.K., and Singh, R.P. (2022). Microbial consortium-based composting: recent trends and innovations. *Waste Manag Res.* 40 (9): 1103–1113.

9 Zhang, Y., Wang, X., Li, Y., et al. (2023). Enhancing aerobic composting with biochar and microbial inoculants. *Sci Total Environ.* 855: 158967.

10 Hu, Z., Ma, L., and Geng, W. (2022). Anaerobic digestion: a promising technology for organic waste valorization. *J Clean Prod.* 365: 132721.

11 Lee, D.H., Kim, S., and Park, J. (2023). Recent progress in anaerobic membrane bioreactors for sustainable biogas production. *Renew Energy.* 201: 150–162.

12 Thakur, P. and Raj, A. (2022). Solid-state fermentation of agro-industrial waste for enzyme production. *Biochem Eng J.* 186: 108567.

13 Kumar, A., Joshi, A., and Jain, R. (2023). Optimization strategies for submerged fermentation in microbial bioprocessing. *Curr Biotechnol.* 12 (1): 24–36.

14 Elmekawy, A., Zaky, A.S., and Hegab, H.M. (2022). Microbial fuel cells: recent advances and potential applications in waste valorization. *Bioresour Technol.* 343: 126077.

15 Sen, S., Saha, P., and Banerjee, A. (2023). Microbial enzymes in waste valorization: novel trends and biotechnological applications. *Environ Sci Pollut Res Int.* 30: 12700–12721.

16 Ghosh, A., Basu, S., and Ghosh, S. (2022). Synthetic microbial consortia: applications in sustainable bioprocessing. *Crit Rev Biotechnol.* 42 (6): 952–968.

17 Ali, M., Singh, H., and Yadav, D. (2023). Precision fermentation in microbial biotechnology: current status and future directions. *Trends Microbiol.* 31 (2): 120–133.

18 Verma, M., Kumar, A., and Das, P. (2023). Microalgae-based phototrophic systems for integrated waste valorization. *Renew Sustain Energy Rev.* 156: 112074.

19 Singh, M., Patil, R., and Joshi, D. (2023). Decentralized composting with microbial consortia: a case from Pune, India. *Waste Manag.* 157: 73–81.

20 Larsen, M., Jensen, K., and Holst, I. (2022). Anaerobic digestion of dairy and food waste in Denmark: co-digestion strategies and microbial insights. *Bioresour Technol Rep.* 20: 101291.

21 Da Silva, M.L., Ferreira, R.A., and Souza, D.O. (2023). Biorefinery valorization of sugarcane residues: a Brazilian case study. *Ind Crops Prod.* 195: 116377.

22 Watanabe, K., Saito, T., and Yoshida, S. (2023). Enzymatic biodegradation of plastic waste using engineered microbial systems. *J Clean Prod.* 405: 136938.

23 Otieno, E.O., Mwangi, H.N., and Kimani, S. (2022). Microbial fuel cells for sustainable sanitation and energy in slum settlements: Kenyan pilot study. *Environ Technol Innov.* 28: 102938.

24 Al Kaabi, J., Farooq, F., and Khan, F. (2022). Microalgae photobioreactors for sustainable greenhouse operations in arid environments. *Sci Total Environ.* 834: 155205.

25 Nolan, R., Cheng, A., and Patel, R. (2023). Upcycling food waste to functional mycoproteins via precision fermentation. *Biotechnol Adv.* 64: 108167.

2

Waste Valorization: A Raw Material Source for Industry

2.1 Introduction

Waste valorization represents a transformative approach to managing industrial, municipal, and agricultural waste, moving away from traditional disposal methods and toward the recovery of valuable resources. At its core, waste valorization refers to the conversion of waste materials into more valuable products such as fuels, chemicals, construction materials, and energy. This concept is central to achieving a sustainable, circular economy by reducing reliance on virgin resources and minimizing environmental degradation.

Valorization is not a singular technique but a broad umbrella encompassing several scientific and engineering processes. These methods are generally categorized into four key types: biological, chemical, thermal, and physical. Biological processes utilize microorganisms or enzymes to convert organic waste into bio-based products like biogas, organic acids, and compost. Chemical processes involve reactions such as depolymerization and transesterification to recover monomers or synthesize fuels and platform chemicals from complex waste streams. Thermal processes, such as pyrolysis and gasification, apply heat to break down waste into energy-rich gases, oils, or char. Lastly, physical processes focus on separation, sorting, and mechanical treatment to recover usable materials or prepare waste for further processing [1].

These diverse methods allow industries to adapt valorization strategies based on the specific characteristics of the waste generated and the desired end products. The scope of waste valorization has expanded significantly in recent years, supported by advancements in biotechnology, materials science, and process engineering. It now plays a crucial role in reducing environmental footprints, supporting green innovation, and creating circular resource loops across sectors ranging from agriculture and energy to manufacturing and construction.

Biological Processes and Techniques in Zero Waste Management, First Edition. Garima Awasthi and Mohit Nigam. Published 2026 by Wiley-VCH GmbH.

By reimagining waste not as an endpoint but as a beginning for resource generation, waste valorization has become a pivotal pillar in the development of resilient, low-impact industrial systems.

The volume of waste generated worldwide is rising at an unprecedented rate, driven by rapid urbanization, population growth, industrial expansion, and consumer behavior. According to the World Bank, global municipal solid waste (MSW) generation has already surpassed 2 billion tonnes annually and is projected to reach nearly 4 billion tonnes by 2050 if current trends continue. This alarming increase places immense pressure on traditional waste management infrastructure, exacerbating issues like landfill overuse, pollution, and greenhouse gas emissions.

At the same time, industries across the globe are experiencing mounting pressure to reduce their environmental impacts and diversify their raw material sources. As the extraction and processing of virgin materials become more expensive and ecologically damaging, the demand for alternative feedstocks has grown substantially. Industries such as packaging, textiles, construction, and electronics are actively seeking renewable, low-impact raw materials to meet sustainability goals and comply with increasingly strict regulatory frameworks.

Waste valorization offers a viable solution to both challenges: it addresses the issue of mounting waste and provides a steady supply of secondary raw materials. For example, food and agricultural residues can be converted into bio-based plastics or biofuels, construction and demolition debris can be reused as aggregates, and e-waste can be mined for rare earth metals and critical minerals. In doing so, valorization not only alleviates the burden of waste disposal but also reduces industrial dependency on finite natural resources.

Furthermore, policy developments such as carbon pricing, landfill taxation, and extended producer responsibility (EPR) schemes have made waste-derived materials more competitive in global markets. In response, industries are increasingly investing in circular solutions and sustainable procurement strategies, further enhancing the relevance and demand for waste valorization.

Traditional industrial systems are built upon a linear model—one that follows a straight line from resource extraction to product manufacture, use, and disposal. This model is inherently unsustainable, as it relies on continuous consumption of finite resources and generates large amounts of waste, often with little or no recovery. The consequences include environmental degradation, resource depletion, and a growing disconnect between production systems and ecological balance [2].

In contrast, the circular economy is founded on the principles of regeneration, reuse, and resource optimization. Instead of discarding materials after a single use, the circular model keeps resources in use for as long as possible, extracting maximum value from them before regenerating products and materials at the end of their life cycle. Waste valorization is a cornerstone of this model, providing the technological and biological means to reincorporate waste into the value chain.

For instance, bio-based industries can use agricultural waste to produce industrial enzymes and biodegradable polymers. The energy sector can convert organic municipal waste into biogas for heat and power generation. Electronics manufacturers can recover critical metals from discarded devices and reintegrate them into new products. These practices not only reduce environmental pressure but also create closed-loop supply chains that enhance material security and economic resilience.

The shift from linear to circular models is not merely an environmental necessity—it is increasingly becoming a strategic and economic imperative. Companies adopting circular practices often benefit from reduced input costs, improved regulatory compliance, increased consumer trust, and access to green financing. Global corporations such as Unilever, IKEA, and Tesla have integrated waste valorization into their operations as part of broader sustainability commitments.

In essence, waste valorization enables industries to transition from resource extraction and disposal toward regeneration and continuous value creation, helping to build more sustainable, innovative, and future-ready industrial systems.

2.2 The Need for Industrial Waste Valorization

The growing urgency to adopt industrial waste valorization stems from a convergence of environmental, regulatory, economic, and strategic imperatives. One of the most significant drivers is the tightening of environmental regulations across the globe. Governments and international organizations have introduced a range of policies aimed at reducing industrial pollution, minimizing waste, and promoting resource efficiency. Frameworks such as the European Union's Waste Framework Directive, EPR mandates, and the ESG (Environmental, Social, and Governance) reporting standards now compel industries to not only manage their waste more responsibly but also demonstrate sustainable practices to regulators, investors, and the public. Alongside these regulatory pressures, the global supply chain faces increasing vulnerability due to the depletion of natural resources and geopolitical instability. The scarcity of essential raw materials, such as rare earth elements, fossil fuels, and fresh water, has been further exacerbated by events like the COVID-19 pandemic and conflicts such as the Russia–Ukraine war, which disrupted international trade and highlighted the fragility of global supply lines. As a result, industries are seeking localized, resilient, and renewable alternatives to conventional resources needs that can be met through the strategic recovery of valuable materials from industrial and municipal waste streams. Furthermore, the rise of the circular economy model has prompted industries to redesign their operations for long-term sustainability and efficiency. Waste valorization plays a critical role in this transition by enabling the reuse and recycling of materials within and across industrial sectors. Concepts such as industrial symbiosis, where the waste from one facility serves as the input for another, are becoming increasingly common, as seen in successful models like the Kalundborg Symbiosis in Denmark. Certifications such as ISO 14001 and LEED are also encouraging companies to integrate waste valorization into their environmental management systems. Collectively, these factors highlight that waste valorization is not just a compliance measure but a strategic approach to reducing environmental impact, mitigating supply chain risks, and building competitive, future-ready industries [3].

2.3 Types of Waste for Industrial Valorization

Industrial waste valorization is most effective when tailored to the specific characteristics and composition of waste streams. Different types of waste ranging from biodegradable organics to complex synthetic and hazardous materials require distinct treatment technologies and conversion pathways. This section discusses the primary waste categories suitable for industrial valorization, highlighting their sources, typical composition, and transformation methods that enable their conversion into useful raw materials.

2.3.1 Organic and Agricultural Waste

Organic and agricultural waste represents one of the most abundant and underutilized resources in the global waste landscape. This category includes biodegradable materials derived from plant or animal sources, often generated in large volumes by farming, food processing, and forestry operations. Common examples are crop residues (e.g., rice straw, wheat husks, corn stalks), fruit and vegetable peels, spent grains, food scraps, livestock manure, and residues from dairy or slaughterhouse operations.

These organic materials are rich in carbon, nitrogen, and other nutrients, making them highly suitable for a range of valorization processes. Composting is a traditional yet effective biological method for transforming organic waste into nutrient-rich compost, which can improve soil fertility and structure when applied in agriculture or landscaping. This process involves aerobic microbial decomposition and is typically low-cost and easy to implement, especially in rural and peri-urban areas [4].

Anaerobic digestion (AD) is another widely adopted technique, where microbial consortia break down organic matter in the absence of oxygen to produce biogas (a mixture of methane and carbon dioxide) and digestate, a nutrient-dense slurry. The biogas can be used for heating, electricity generation, or upgraded to biomethane for grid injection or vehicle fuel. The digestate, after stabilization, serves as an organic fertilizer.

Fermentation processes, including solid-state and submerged fermentation, are used to derive value-added products such as ethanol, lactic acid, enzymes, and amino acids. For example, sugar-rich fruit waste can be fermented into bioethanol, while oil-rich residues can support the growth of microorganisms producing biosurfactants or biopolymers.

Pyrolysis offers a thermochemical route for converting lignocellulosic agricultural residues into bio-oil, syngas, and biochar. Bio-oil can be refined into fuels or specialty chemicals, while biochar can enhance soil carbon content and sequester greenhouse gases. These pathways not only mitigate environmental issues associated with open burning or uncontrolled decomposition of organic waste but also support rural livelihoods through bioeconomy development.

2.3.2 Industrial and Manufacturing Waste

Industrial and manufacturing operations generate large volumes of solid and liquid waste, often containing valuable components such as metals, heat energy, and residual chemicals. These wastes originate from diverse sectors including metal processing, textiles, cement production, ceramics, paper and pulp, and chemical manufacturing.

A key strategy for valorizing such waste is metal recovery. Industrial sludges, slags, and dusts may contain significant quantities of recoverable metals such as copper, zinc, iron, or aluminum. Technologies like hydrometallurgy (acid or solvent leaching), pyrometallurgy, and biosorption (using microbial or biochar-based systems) are used to extract and purify these metals. Recovery not only reduces the need for mining virgin ores but also prevents the release of toxic metals into the environment.

Another important area is catalyst recovery and regeneration. Spent catalysts from petrochemical and pharmaceutical industries, often composed of costly metals like platinum or palladium, can be regenerated through physical or chemical treatment to restore activity. In some cases, valuable metals are completely recovered and recycled into new catalyst formulations.

Industrial waste also includes fly ash, bottom ash, and slag, which can be valorized as supplementary cementitious materials (SCMs) in concrete production. Their incorporation into construction materials not only diverts large quantities of waste from landfills but also reduces the carbon footprint of cement, a major industrial emitter of greenhouse gases [5].

In addition, waste heat from high-temperature industrial operations can be captured using thermoelectric generators or heat exchangers to generate electricity or preheat input streams, thereby improving overall energy efficiency.

By adopting waste valorization strategies, manufacturing industries can reduce their raw material consumption, lower disposal costs, and meet increasingly stringent environmental regulations, all while creating new value streams from previously discarded by-products.

2.3.3 Electronic and Hazardous Waste

Electronic waste, or e-waste, is among the fastest-growing waste streams globally, driven by rapid technological advancements and shortened product lifespans. E-waste includes discarded consumer electronics such as smartphones, computers, televisions, batteries, and other devices. These products contain a complex mixture of valuable and hazardous materials, including precious metals (gold, silver, palladium), rare earth elements, plastics, and toxic substances such as lead, mercury, and brominated flame retardants.

One of the most promising valorization techniques for e-waste is urban mining—a process of recovering valuable metals from used electronics. Conventional methods involve mechanical shredding and sorting, followed by chemical leaching to isolate specific elements. More sustainable alternatives such as bioleaching use metal-tolerant microorganisms like *Acidithiobacillus ferrooxidans* to extract metals from circuit boards and other components under controlled conditions.

Beyond metals, plastics and glass from electronic devices can be recovered and reused in the production of insulation materials, tiles, and composites. Solvent-based extraction and supercritical fluid processes are being developed to separate and purify specific polymer fractions for reuse in high-value applications [6].

Proper handling of hazardous fractions is critical, as improper disposal of e-waste can lead to soil and water contamination and pose serious health risks. Thus, e-waste valorization not only supports the circular economy but also addresses critical environmental and public health concerns, especially in regions where informal recycling practices are prevalent.

2.3.4 MSW and Plastics

MSW encompasses a diverse range of materials generated by households, businesses, and institutions, including organic waste, paper, plastics, textiles, glass, and metals. Of particular concern is plastic waste, much of which is nonbiodegradable and accumulates in landfills and oceans, contributing to long-term pollution.

Plastic valorization can be achieved through mechanical recycling, where plastic waste is cleaned, shredded, and remolded into new products. This method is most effective for single-polymer plastics like Polyethylene Terephthalate (PET) and High-Density Polyethylene (HDPE). However, contaminated or mixed plastic streams often require more advanced treatments.

Chemical recycling techniques, such as pyrolysis, gasification, and depolymerization, break down polymers into their monomers or simpler hydrocarbons. These can be reused as raw materials in the production of virgin-quality plastics or fuels. Enzymatic depolymerization, though still in developmental stages, holds promise for selectively degrading specific plastic types like PET under mild conditions.

In the case of biodegradable and organic fractions of MSW, AD and composting can be applied to produce biogas and organic fertilizer, respectively. Solid-state fermentation is also being explored for producing enzymes or organic acids from organic MSW [7].

Emerging innovations include microbial consortia capable of degrading biodegradable plastics and smart sorting systems using artificial intelligence and optical sensors to improve the quality of waste stream segregation at material recovery facilities (MRFs).

Overall, the valorization of MSW and plastics presents both challenges and opportunities. While heterogeneity and contamination complicate processing, successful implementation of integrated waste management strategies can significantly reduce landfill volumes, recover resources, and reduce the environmental footprint of urban living.

2.4 Valorization Techniques and Pathways

The transformation of waste into valuable resources requires a wide array of technological approaches, each tailored to specific waste types and desired end-products. These valorization techniques can be grouped into four primary categories: bioconversion and microbial processes, thermochemical methods, chemical recycling and catalytic conversion, and physical sorting and separation. Together, they enable the conversion of industrial, agricultural, municipal, and hazardous wastes into energy, materials, and chemical feedstocks. Understanding the underlying mechanisms and industrial applications of these pathways is essential for designing efficient and sustainable waste management systems.

2.4.1 Bioconversion and Microbial Processes

Bioconversion refers to the use of biological systems—mainly microorganisms or their enzymes to degrade and convert organic waste into useful products. This approach is particularly effective for treating biodegradable materials such as food waste, agricultural residues, animal waste, and some industrial effluents [8].

2.4.1.1 Anaerobic Digestion

AD is a widely used technique where organic matter is decomposed by microorganisms in the absence of oxygen. This process generates two primary products: **biogas**, which contains methane and carbon dioxide and can be used for energy production, and **digestate**, a nutrient-rich slurry that serves as organic fertilizer. AD systems are employed in wastewater treatment plants, agricultural farms, and food processing industries to manage organic waste while producing renewable energy.

Composting: Composting is an aerobic process that involves the microbial degradation of organic matter in the presence of oxygen. The result is a stable, humus-like material known as compost, which is rich in nutrients and beneficial for soil health. Composting can be carried out on small or large scales and is an effective solution for managing green waste, kitchen scraps, and agricultural residues. It reduces landfill dependence and enhances circularity in agriculture.

Fermentation: Fermentation processes harness specific microorganisms to convert organic substrates, such as sugars and starches, into bio-based chemicals, fuels, or value-added products. For instance, food waste or molasses can be fermented by yeast or bacteria to produce **bioethanol**, **lactic acid**, **butanol**, or **amino acids**. Depending on the end product, fermentation can be performed under aerobic or anaerobic conditions using submerged or solid-state systems. This technique is essential in industries seeking to replace petrochemical inputs with bio-based alternatives.

Bioleaching: Bioleaching is a process where metal-tolerant microorganisms are used to solubilize metals from solid waste, particularly electronic waste or mining residues. Organisms such as *A. ferrooxidans* oxidize sulfides and mobilize metals like copper, zinc, gold, and cobalt into solution. This environmentally friendly method reduces the need for harsh chemical treatments and is increasingly used in the recovery of valuable elements from complex waste matrices [9].

The flexibility of bioconversion makes it an ideal approach for localized and decentralized waste management systems, particularly in rural and peri-urban settings.

2.4.2 Thermochemical Methods

Thermochemical valorization involves the application of heat to transform waste materials into usable energy carriers, fuels, or chemicals. These methods are particularly suited for carbon-rich waste streams such as biomass, plastics, and mixed solid waste.

Pyrolysis: Pyrolysis is the thermal decomposition of organic materials in the absence of oxygen, typically at temperatures between 300 and 600 °C. The process yields three main products: **bio-oil**, which can be refined into fuel or chemicals; **syngas** (a mixture of hydrogen and carbon monoxide); and **biochar**, a carbon-rich solid that can be used as a soil amendment or carbon sequestration agent. Pyrolysis is gaining popularity for converting waste plastics and agricultural biomass into valuable intermediates for energy and materials production.

Gasification: Gasification involves the partial oxidation of carbonaceous waste at high temperatures (700–1500 °C) to produce **syngas**, which can be used for heat, electricity, or as a feedstock in chemical synthesis (e.g., methanol or synthetic diesel). This process is effective for low-moisture wastes such as wood chips, MSW, or coal by-products. Unlike combustion, gasification offers cleaner emissions and higher energy efficiency.

2.4.2.1 Incineration with Energy Recovery

Incineration involves the complete combustion of waste in the presence of oxygen. Modern waste-to-energy plants capture the heat released during combustion to generate steam, electricity, or district heating. While incineration effectively reduces waste volume, it must be carefully managed to minimize the release of pollutants such as dioxins and heavy metals. Advanced emission control systems have made this technique more environmentally acceptable, especially for treating nonrecyclable municipal waste.

These thermochemical pathways are particularly advantageous for high-volume waste treatment and integrated energy recovery systems, making them suitable for urban and industrial applications [10].

2.4.3 Chemical Recycling and Catalytic Conversion

Chemical recycling focuses on breaking down complex waste materials into their chemical constituents so they can be reused as raw materials. Unlike mechanical recycling, which may degrade material quality, chemical processes enable the regeneration of virgin-quality inputs.

Depolymerization: Depolymerization involves chemically breaking down polymers (e.g., PET, PU, PLA) into their original monomers or intermediates. For instance, PET can be depolymerized into terephthalic acid and ethylene glycol, which can then be used to manufacture new plastic products. Techniques like hydrolysis, glycolysis, and methanolysis are used depending on the polymer type.

This method addresses the limitations of mechanical plastic recycling by handling mixed or contaminated plastic waste.

Transesterification: This process is primarily used to produce **biodiesel** from oils and fats. In transesterification, triglycerides react with an alcohol (usually methanol) in the presence of a catalyst (alkaline or acidic) to produce **fatty acid methyl esters (FAMEs)** and glycerol. Waste cooking oil, animal fat, and oil-rich algae are common feedstocks. This approach offers a sustainable route for transforming lipid-based waste into renewable fuels [11].

2.4.3.1 Hydrothermal Liquefaction

HTL involves the treatment of wet biomass in water at elevated temperature (250–370 °C) and pressure (10–25 MPa). Under these conditions, the biomass breaks down into **bio-crude**, a substance similar to petroleum crude that can be further refined. Unlike pyrolysis or gasification, HTL does not require the biomass to be dried, making it suitable for high-moisture wastes such as algae, sewage sludge, and food waste.

Catalytic Conversion methods, including catalytic cracking and reforming, are also used to upgrade waste-derived oils into high-value fuels and chemicals. These methods are essential for creating a closed-loop material cycle, especially in the plastics and petrochemical industries.

2.4.4 Physical Sorting, Separation, and Processing

Physical valorization techniques are critical in preparing waste materials for further chemical or biological treatment. These methods involve sorting, separation, and mechanical treatment to enhance the purity, quality, and usability of recyclable materials [12].

2.4.4.1 Mechanical Shredding and Grinding

These processes reduce the size of waste materials, increasing their surface area for subsequent processing steps. Shredding is commonly used for plastics, paper, wood, and e-waste to facilitate easier handling and improve reaction efficiency in thermal or biochemical treatments.

2.4.4.2 Magnetic and Eddy Current Separation

These techniques are used to recover metallic components from mixed waste. Magnetic separation removes ferrous metals, while eddy current systems separate nonferrous metals like aluminum and copper. These methods are essential in MRFs and e-waste recycling operations.

2.4.4.3 Air Classification and Density Separation

These systems separate materials based on their weight and aerodynamic properties. Light materials (e.g., paper, plastic films) are separated from heavier ones (e.g., glass, metals), improving sorting efficiency and product quality.

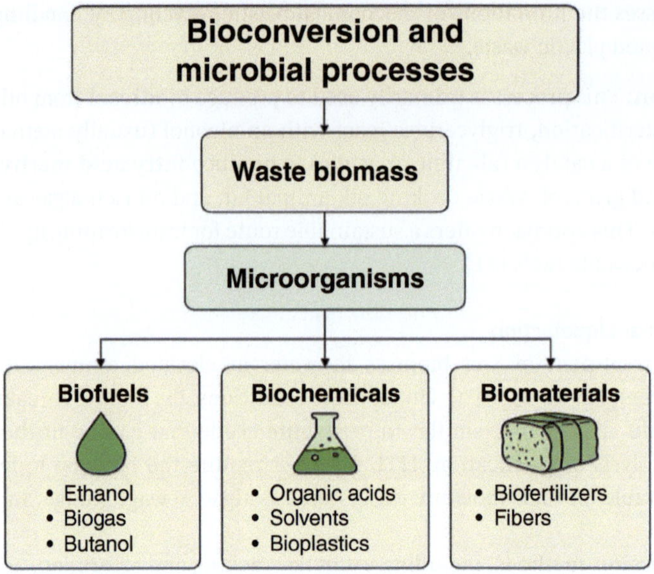

Figure 2.1 Flowchart illustrating the bioconversion of waste biomass.

2.4.4.4 Membrane Filtration and Ultrafiltration

Used primarily in liquid waste treatment, membrane systems separate dissolved or suspended solids from water or valuable solutes. Applications include wastewater reclamation, enzyme recovery, and nutrient extraction from digestate or industrial effluents.

2.4.4.5 Optical and Robotic Sorters

Advanced facilities now use optical recognition systems, AI, and robotics to automate waste sorting. These technologies improve efficiency, reduce labor costs, and enhance the purity of recovered fractions, especially in complex waste streams like mixed plastics.

Physical techniques form the foundation of integrated waste management systems and significantly influence the performance of downstream valorization processes (Figure 2.1).

2.5 Industrial Applications of Waste-derived Raw Materials

The transformation of waste into valuable industrial inputs has opened new avenues for sustainable manufacturing across various sectors. Among the most prominent applications is the production of bioplastics and biopolymers from organic waste streams. Materials such as polyhydroxyalkanoates (PHA) and polylactic acid (PLA) are synthesized by microbial fermentation of sugar- or starch-rich agricultural residues like potato peels, corn husks, or food scraps. These biodegradable alternatives to conventional plastics are increasingly used in packaging, medical devices, and consumer goods. Starch-based bioplastics, derived from waste biomass, offer further potential for disposable items and compostable products, significantly reducing the environmental footprint of plastic use.

Another impactful application lies in the construction industry through the development of green building materials sourced from industrial and municipal waste. For instance, fly ash from thermal power plants and slag from steel production can replace a portion of cement in concrete, enhancing strength while reducing carbon emissions associated with clinker production. Additionally, recycled aggregates obtained from construction and demolition waste are used in road base layers, drainage systems, and new concrete, promoting circularity in infrastructure projects. Innovative solutions like bio-bricks, produced using microbial-induced calcite precipitation, utilize agricultural waste and soil, providing an eco-friendly alternative to fired clay bricks with lower energy requirements [13].

Waste valorization also plays a key role in the generation of renewable energy and fuels. Biogas, produced via AD of food waste, livestock manure, and sewage sludge, is a clean and efficient energy carrier that can be used for heating, electricity, or as vehicle fuel after purification. Similarly, biodiesel is synthesized from waste cooking oils and animal fats through transesterification and serves as a substitute for petroleum diesel in transportation. Bioethanol, obtained from lignocellulosic waste through fermentation, is widely used as a bio-blend fuel. Additionally, biochar, a carbon-rich by-product of pyrolysis, not only sequesters carbon but also improves soil fertility, making it a valuable soil amendment in agriculture.

In the chemical and nutraceutical sectors, waste-derived substrates are increasingly utilized for the production of high-value platform chemicals and health-related compounds. Organic waste streams are fermented to yield lactic acid, which serves in the production of bioplastics and food preservatives. Succinic acid and citric acid, both platform chemicals for solvents, resins, and pharmaceuticals, are efficiently produced using microbial cultures on agro-industrial waste. Furthermore, omega-3 fatty acids, with growing demand in the health and supplement industry, are being sourced from microalgae cultivated on wastewater or carbon dioxide streams.

Finally, the integration of waste valorization into circular manufacturing and eco-design is reshaping industrial practices. Sectors such as textiles are recycling fibers from waste garments, while electronics manufacturers are recovering rare earth elements and metals from e-waste for reuse in new devices. In agriculture, crop residues are being transformed into biodegradable packaging or mulch films, which naturally degrade back into the soil. These innovations underscore the shift from linear production to systems that prioritize material recovery, resource efficiency, and long-term environmental sustainability [14].

2.6 Technological Innovations and Scaling Up

The advancement and scalability of waste valorization processes are being driven by rapid developments in technology, particularly in biotechnology, engineering, digital systems, and industrial integration. One of the most transformative areas is biotechnological innovation, where synthetic biology and engineered microorganisms are being developed to efficiently convert specific waste substrates into high-value products. Through gene editing tools such as CRISPR, microbes like *Escherichia coli* and *Pseudomonas putida* are tailored to tolerate toxic compounds and produce targeted chemicals, fuels, or bioplastics. Moreover, microbial consortia, co-cultures of different microbial species—are being designed to work synergistically for the breakdown of complex waste materials, such as lignocellulose or mixed food waste. These consortia can metabolize diverse components simultaneously, leading to more complete and efficient conversion pathways.

In parallel, progress in process engineering and reactor technologies is enabling the scale-up of laboratory methods to pilot and industrial levels. Modular bioreactors are now being developed with plug-and-play capabilities, allowing for flexible deployment across various locations and waste types. These systems can be customized or scaled based on feedstock availability and product demand. Additionally, hybrid systems that combine biological and thermochemical stages, such as pyrolysis followed by microbial fermentation—offer improved efficiency and multiproduct recovery from a single waste stream. Such integrated designs maximize carbon and energy recovery while minimizing environmental impact [15].

Another critical area supporting waste valorization at scale is smart separation and recovery technologies. Efficient downstream processing is essential to ensure the purity and usability of valorized products. Membrane bioreactors (MBRs) are widely used for treating wastewater and separating valuable compounds such as organic acids or nutrients. Similarly, adsorption-based systems using materials like activated carbon, biochar, or nanomaterials help capture metals, dyes, and other pollutants. Advanced solvent extraction systems are also employed to selectively isolate chemicals or metals from complex waste streams, particularly in e-waste or biomass processing.

Digital technologies have further enhanced the operational control and optimization of waste valorization systems. The integration of data analytics, artificial intelligence (AI), and digital twins enables real-time monitoring and predictive control. IoT-enabled sensors continuously track parameters like pH, temperature, and gas composition, feeding data into AI models that adjust conditions automatically to optimize yield and process stability. Digital twins, or virtual replicas of physical systems, allow engineers to simulate different operational scenarios before implementation. This reduces risk, lowers costs, and shortens development timelines during scale-up. Finally, the successful deployment of these technologies depends on circular ecosystem integration and industrial symbiosis. In eco-industrial parks, multiple industries co-locate and share resources such as waste, energy, water, and infrastructure. For example, waste heat from one facility can be used in another, or one company's by-products can serve as raw materials for a neighboring enterprise. This interconnected approach enhances resource efficiency, reduces emissions, and supports regional economic development through shared value creation and innovation in waste reuse [16].

2.7 Policy Outlook

2.7.1 Enabling Policy Ecosystems

The role of government in promoting waste valorization is critical for creating an enabling environment that supports sustainable industry transformation. Policymakers can drive adoption by implementing legislation that incentivizes resource efficiency, mandates waste reduction, and provides financial or infrastructural support for valorization technologies. National and international frameworks, such as the European Union's Circular Economy Action Plan, India's GOBAR-Dhan Scheme, and the United Nations Sustainable Development Goals (SDGs), serve as guiding structures for integrating waste valorization into broader environmental and economic development strategies. By aligning local policy with global sustainability goals, governments can encourage industries to adopt innovative waste-to-resource solutions.

2.7.2 Market and Investment Readiness

To scale waste valorization beyond pilot projects, robust investment mechanisms are essential. Public–private partnerships (PPPs), green bonds, and blended finance models can bridge the funding gap for early-stage technologies. Moreover, ESG considerations are increasingly influencing investor behavior. Companies that demonstrate alignment with ESG criteria, through transparent reporting, lifecycle assessments, and sustainable supply chains, are more likely to attract ethical and long-term capital. Financial instruments such as feed-in tariffs for renewable energy or tax credits for bio-based materials can further improve the commercial viability of valorization initiatives and attract market participation [17].

2.7.3 International Collaboration and Harmonization

Given that waste generation, resource scarcity, and pollution are global challenges, international cooperation is vital. Harmonized standards for waste classification, product certification, and valorization technologies can facilitate global trade in waste-derived materials and improve regulatory compliance across borders. Organizations like the Global Waste Valorization Alliance (GWVA), the Basel Convention, and UNIDO play key roles in knowledge sharing, technology transfer, and capacity building. Cross-border collaboration also ensures that best practices in waste management and circular economy implementation are widely disseminated and adapted to local contexts.

2.7.4 Public Engagement and Awareness

Public perception significantly influences the success of waste valorization initiatives. Many consumers remain skeptical about using products derived from waste, especially in sectors like packaging, cosmetics, and food contact materials. To address this, governments and industries must invest in education, community outreach, and eco-labeling programs that transparently communicate the safety, quality, and sustainability benefits of waste-derived products. Including waste valorization in school curricula, promoting sustainability campaigns, and offering incentives for eco-conscious consumer behavior can contribute to greater acceptance and cultural normalization of circular practices [18].

2.7.5 Future Vision: Waste-free Industries

The long-term vision of waste valorization extends beyond resource recovery, it envisions the complete transformation of industrial systems into waste-free, regenerative ecosystems. This includes adopting design-for-reuse principles, where products are created with disassembly and recyclability in mind, and implementing closed-loop manufacturing systems that continuously cycle materials. Emerging technologies such as blockchain and IoT-based tracking can help monitor material flows in real time, ensuring transparency and traceability throughout the product life cycle. Future industries will likely rely on integrated platforms that connect suppliers, processors, and consumers in a circular network, redefining waste not as a burden but as a valuable resource at every stage.

2.8 Economic Feasibility and Market Dynamics

The long-term success of industrial waste valorization hinges not only on technological efficiency but also on economic viability. As industries shift toward circular economy models, understanding the cost-effectiveness and market dynamics of waste valorization technologies becomes critical. This section evaluates the economic feasibility of these systems through cost-benefit assessments, analyzes the evolving market demand for waste-derived products, and explores investment mechanisms and policy frameworks that enable scalable implementation while managing financial risk [19].

2.8.1 Cost-benefit Analysis of Valorization Technologies

A comprehensive cost-benefit analysis (CBA) is essential for determining whether waste valorization technologies are commercially viable at pilot or industrial scale. The two core cost components of any valorization project are Capital Expenditure (CAPEX) and Operating Expenditure (OPEX). CAPEX includes the initial investment required for setting up facilities, purchasing equipment (such as reactors, separators, and monitoring systems), and building infrastructure. In contrast, OPEX refers to recurring costs such as energy consumption, labor, feedstock collection and preprocessing, maintenance, and waste disposal.

When assessing the return on investment, key economic tools such as Net Present Value (NPV) and Internal Rate of Return (IRR) are commonly employed. NPV calculates the value of projected cash flows over time, discounted to present value, allowing stakeholders to determine whether the investment yields a net gain. A positive NPV indicates profitability. IRR, on the other hand, measures the rate at which investment returns match or exceed initial costs, providing a benchmark for comparing different projects. A higher IRR typically signifies a more attractive investment opportunity. Other commonly used tools include Payback Period (PBP) and Life Cycle Costing (LCC), which account for long-term cost implications, including environmental externalities.

While the economic performance of valorization projects varies by feedstock and technology, certain pathways, such as AD for biogas, fermentation for bioethanol, and pyrolysis for biochar, have demonstrated reasonable returns on investment within 4–6 years, especially when supported by government subsidies or carbon credit schemes. In high-value waste streams like e-waste, metal recovery through bioleaching or hydrometallurgy can yield significant financial returns, especially with efficient recovery rates and scale.

Additionally, hybrid systems that combine biological and thermochemical processes may involve higher upfront costs but offer multiple revenue streams (e.g., biofuels, chemicals, electricity), enhancing their overall economic sustainability. Therefore, careful techno-economic modeling is critical for project developers, investors, and policymakers to assess the feasibility and scalability of waste valorization projects [20].

2.8.2 Market Demand for Waste-derived Materials

The global push for sustainability has created a growing market for products derived from waste, with sectors such as packaging, energy, agriculture, electronics, and construction demonstrating increasing interest in renewable, circular alternatives. Bioplastics, particularly PHA and PLA derived from organic waste, have witnessed a significant surge in demand. The global bioplastics market is projected to grow at a compound annual growth rate (CAGR) of over 14%, reaching a valuation of nearly USD 29 billion by 2028. This growth is driven by rising environmental regulations banning single-use plastics, as well as consumer preference for biodegradable packaging materials.

Similarly, biofuels derived from organic and industrial waste streams, including biodiesel, biogas, and bioethanol, are gaining traction as cleaner alternatives to fossil fuels. Governments across the globe are implementing blending mandates and incentivizing the use of second-generation biofuels, which are derived from non-food biomass such as agricultural residues or MSW. According to the International Energy Agency (IEA), waste-based biofuels could account for 25% of global biofuel production by 2030, indicating a strong upward trend in demand.

Another sector showing rapid growth is e-waste recycling. With the increasing adoption of electronics and electric vehicles (EVs), demand for critical raw materials such as lithium, cobalt, palladium, and rare earth elements is surging. Valorization of electronic waste through urban mining and microbial metal recovery offers a viable route to address this demand while reducing reliance on virgin mineral extraction. The e-waste recycling market is estimated to reach over USD 100 billion in the coming decade, driven by both environmental regulations and supply chain pressures.

In construction, recycled materials such as fly ash, slag, and crushed concrete are replacing virgin aggregates and cement additives, supporting the development of green buildings and infrastructure. In agriculture, by-products like compost, digestate, and biochar are being adopted as organic soil amendments, meeting the growing demand for sustainable farming inputs.

Overall, waste-derived materials are no longer niche alternatives, they are becoming mainstream commodities with significant commercial potential. Their adoption is being driven not only by environmental concerns but also by regulatory mandates, supply chain disruptions, and a growing preference for traceable and low-carbon materials [21].

2.8.3 Investment, Policy Incentives, and Risk Management

Despite the promising market outlook, large-scale deployment of waste valorization technologies requires substantial investment and a supportive policy ecosystem. PPPs have emerged as a successful model for financing such projects, wherein governments provide land, infrastructure, or grants while private entities manage operations and innovation. These partnerships distribute risk and ensure shared accountability, particularly in capital-intensive projects like waste-to-energy plants or biorefineries.

Blended finance is another strategic approach that combines public funds (grants, concessional loans) with private investment to reduce perceived risk and attract commercial capital. International financial institutions and development banks often provide seed funding, technical assistance, or credit guarantees to initiate waste valorization projects in emerging economies.

Additionally, green bonds and impact investing have gained popularity as tools to channel funds into environmentally beneficial ventures. Green bonds offer fixed returns for investors and are tied to verified environmental outcomes, while impact investors seek both financial and social/environmental returns. Companies that can demonstrate measurable ESG performance such as reduced carbon emissions, improved resource efficiency, or community benefits are increasingly eligible for such funding. To support these financing mechanisms, governments must offer policy incentives such as tax relief, renewable energy feed-in tariffs, waste disposal levies, or priority procurement of recycled products. For example, landfill taxes and carbon credits can make valorization more cost-effective than traditional waste disposal.

Nevertheless, financial risks remain, particularly in the form of feedstock variability, market price fluctuations, and technological uncertainties at scale. Effective risk management strategies include long-term contracts with waste suppliers, diversified revenue models, scalable modular technologies, and insurance schemes that protect against policy reversals or operational disruptions. Sensitivity analysis, scenario modeling, and lifecycle assessments also play key roles in evaluating economic resilience and environmental benefits [22].

In essence, de-risking waste valorization projects through smart investment structures, robust policy support, and transparent governance can unlock the full potential of circular industries and attract sustained investor interest.

2.9 Case Studies and Global Practices

The successful implementation of waste valorization across the globe offers valuable insights into how different sectors industry, corporate enterprises, and governments are leveraging innovation, collaboration, and policy support to transform waste into a valuable resource. The following case studies highlight practical models that have demonstrated both environmental and economic benefits.

2.9.1 Industrial Symbiosis Parks

One of the most celebrated examples of industrial symbiosis is the Kalundborg Symbiosis in Denmark. This eco-industrial park began in the 1960s and has since evolved into a benchmark for circular industrial collaboration. In Kalundborg, multiple industries include a power plant, pharmaceutical manufacturer, oil refinery, and municipal utility exchange waste, energy, and water. For example, excess heat from the power station is used to heat nearby homes and supply steam to adjacent industries. Fly ash from power generation is repurposed for cement production, and wastewater is treated and reused across various facilities. This integrated system has resulted in significant reductions in CO_2 emissions, water consumption, and material waste while enhancing economic resilience and reducing operational costs for all participants.

Similarly, the Tianjin Economic-Technological Development Area (TEDA) in China showcases how planned eco-industrial development can promote waste valorization on a regional scale. The park hosts over 30 enterprises engaged in the exchange of by-products, energy, and resources. For instance, food processing waste is converted into biogas and organic fertilizer, while paper mill residues are processed into value-added chemicals. Government support in the form of zoning regulations, subsidies, and infrastructure has been a key to TEDA's success, highlighting the importance of integrated planning and public–private cooperation in enabling circular practices [23].

2.9.2 Corporate Initiatives in Waste-to-resource Programs

Major corporations are increasingly adopting circular economy strategies to reduce their environmental footprint and align with sustainability goals. Unilever, for example, has implemented a global "Zero Waste to Landfill" program across its manufacturing sites. Through resource recovery, composting, and recycling initiatives, Unilever has diverted over 99% of its nonhazardous waste away from landfills. Organic waste is often valorized into biogas or compost, while plastic and paper waste are reused or converted into fuel. These efforts have not only saved operational costs but also improved the company's ESG performance.

IKEA has committed to becoming a fully circular business by 2030. The company uses recycled wood, textiles, and plastics in many of its furniture products and has established take-back and refurbishment programs across several markets. In a landmark move, IKEA also introduced a line of home furnishings made from ocean plastic and postconsumer waste, showcasing how design innovation can enable large-scale waste valorization.

Braskem, a Brazilian petrochemical giant, leads in the production of bio-based plastics. Its "I'm green™" polyethylene is made from sugarcane bagasse, a by-product of ethanol production. This renewable plastic has a lower carbon footprint compared to conventional fossil-based alternatives and is used by brands like LEGO and Danone. Braskem's model exemplifies how agricultural waste can be integrated into high-value manufacturing supply chains [24].

2.9.3 Government-supported Valorization Projects

Governments worldwide have initiated programs to support the development and deployment of waste valorization technologies. In India, the GOBAR-Dhan (Galvanizing Organic Bio-Agro Resources Dhan) scheme promotes the installation of biogas and composting units in rural areas. The initiative encourages the use of animal dung and agricultural waste for bioenergy and organic fertilizer production. Hundreds of community-level biogas plants have been established, generating clean energy for cooking and farming needs while improving rural sanitation and income opportunities.

The European Union's LIFE Programme is one of the largest environmental funding mechanisms in the world. It supports pilot and demonstration projects across member states focused on waste reduction, resource recovery, and sustainable production. Projects under LIFE include converting pulp and paper waste into platform chemicals and using enzymatic processes to depolymerize textile waste. These initiatives have accelerated technological innovation and cross-border collaboration within the EU. In the United States, the Bioenergy Technologies Office (BETO) under the Department of Energy funds research and commercialization of biofuels and bio-based products derived from waste. BETO supports projects involving MSW, algae cultivation, and enzyme-based plastic degradation. The program not only advances technology but also promotes workforce development and green job creation across the country [25].

2.10 Conclusion

Waste valorization has emerged as a vital solution for addressing global environmental, economic, and resource challenges. By converting industrial, agricultural, municipal, and hazardous waste into valuable products, it enables industries to reduce pollution, conserve natural resources, and align with circular economy goals. Technological innovations ranging from synthetic biology and thermochemical processing to AI-driven reactors and smart separation systems have significantly enhanced the feasibility and scalability of valorization pathways. However, the widespread adoption of these technologies requires supportive policies, informed consumers, and well-coordinated market mechanisms. Governments must provide the regulatory framework, financial incentives, and infrastructure needed to foster innovation and integration. Investors and industries, in turn, must align with ESG principles and embrace sustainable supply chains. Public engagement, international cooperation, and a shared vision for zero-waste, circular economies will be essential for scaling impact. Looking forward, waste valorization is not merely a transitional strategy, it is a foundational component of a resilient, resource-efficient, and sustainable industrial future.

References

1 Khoshnevisan, B., Tsapekos, P., Alfaro, N., and Angelidaki, I. (2022). Waste biorefinery: future strategic framework for sustainable biorefinery development. *Bioresour Technol.* 344: 126195.

2 Mohan, S.V., Modestra, J.A., Amulya, K., et al. (2021). A circular bioeconomy with biobased products from microbial processes: techno-economic and life cycle assessment. *Bioresour Technol.* 319: 124199.

3 Tripathi, N., Hills, C.D., Singh, R.S., and Atkinson, C.J. (2016). Biomass waste utilisation in low-carbon products: harnessing a major potential resource. *Sci Total Environ.* 548–549: 463–471.

4 Ghosh, S.K. (2021). *Waste Valorization and Recycling: Challenges and Opportunities.* Springer.

5 Yuan, H., Wang, X., and Zhang, Y. (2023). Advances in microbial consortia for waste valorization. *J Environ Manage.* 337: 117648.

6 Singh, V., Bansal, S., and Srivastava, A. (2023). Engineered microbial platforms for waste-to-value biotransformation. *Bioresour Technol.* 373: 128721.

7 Khandelwal, H., Dhar, H., Thalla, A.K., and Kumar, S. (2019). Application of life cycle assessment in municipal solid waste management: a worldwide critical review. *J Clean Prod.* 209: 630–654.

8 López-González, D., Fernandez-Lopez, M., Valverde, J.L., and Sanchez-Silva, L. (2013). Pyrolysis of municipal solid waste: influence of char in the syngas composition. *Waste Manage.* 33 (3): 672–678.

9 Zhang, R., Ma, J., and Li Y. (2023). Smart biorefineries using IoT and AI: real-time monitoring and control. *Biotechnol Adv.* 65: 108132.

10 Kumar, A., Chen, H., and Chen, G. (2024). Hybrid bioreactors for organic waste valorization: a review. *Renew Sust Energ Rev.* 172: 113163.

11 Ellen MacArthur Foundation (2019). *Completing the Picture: How the Circular Economy Tackles Climate Change.* UK: Ellen MacArthur Foundation.

12 Lopes, D., Marques, A., and Carvalho, M. (2022). Digital twin technology in industrial bioprocess scale-up. *Comput Chem Eng.* 161: 107772.

13 Basak, B., Ghosh, A., and Pal, P. (2022). Chemical recycling and upcycling of polymer waste: recent developments. *J Environ Chem Eng.* 10 (1): 106901.

14 Nanda, S. and Berruti, F. (2021). Thermochemical conversion of plastic waste to fuels: a review. *Environ Chem Lett.* 19: 123–148.

15 Gao, N., Li, A., and Quan, C. (2020). Pyrolysis and thermal degradation of plastics: from fundamental study to industrial application. *Prog Energy Combust Sci.* 77: 100806.

16 Global Waste Valorization Alliance (GWVA) (2024). Global State of Industrial Waste Reuse 2024. GWVA technical report.

17 United Nations Industrial Development Organization (UNIDO) (2023). *Industrial Waste Valorization Strategies in Emerging Economies.* Vienna: UNIDO.

18 European Commission (2022). *Circular Economy Action Plan: For a Cleaner and More Competitive Europe.* Brussels: EU Publications Office.

19 Kumar, A. and Singh, P. (2023). Financing circular transitions in industry: waste-to-resource pathways. *Sustain Ind Rev.* 5 (2): 45–57.

20 International Energy Agency (IEA) (2023). Renewables 2023: analysis and forecast to 2028. Paris: IEA.

21 Plastics Europe (2022). *Plastics – The Facts 2022: An analysis of European Plastics Production, Demand and Waste Data.* Brussels: Plastics Europe.

22 Unilever (2023). *Making Sustainable Living Commonplace: Sustainability Performance Report.* London: Unilever.

23 IKEA (2022). *People & Planet Positive: Sustainability Strategy Update.* Netherlands: Inter IKEA Systems B.V.

24 Braskem (2023). *I'm Green™ Polyethylene: Renewable Solutions from Sugarcane.* São Paulo: Braskem.

25 Department of Biotechnology (DBT), Government of India (2023). *GOBAR-Dhan: Guidelines and Operational Framework.* New Delhi: DBT.

3

Instrumentation and Control Systems for Biological Valorization of Waste

3.1 Introduction

Biological valorization of waste involves separation of heavy particles and transforming organic waste into valuable products such as bioenergy, biofertilizers, bioplastics, and biomolecules using microbial or enzymatic processes. Various instruments are required to perform these processes. Effective instrumentation and control systems are essential to optimize these processes, ensuring stability, efficiency, and scalability. Converting waste into zero waste involves maximizing reuse, recycling, composting, and energy recovery while minimizing disposal. Instrumentation plays a crucial role in enabling and monitoring this transformation by ensuring that materials are properly sorted, treated, and repurposed. Instrumentation and control systems play a vital role in enhancing the efficiency of biological waste valorization by ensuring precise management of critical process parameters (CPPs). These systems continuously track and regulate factors such as temperature, pH, nutrient concentration, and gas emissions, particularly biogas in anaerobic digestion, to maintain ideal conditions for microbial metabolism and bioproduct synthesis. Integration of advanced sensors, automated data acquisition tools, and intelligent control algorithms allows for real-time monitoring and responsive adjustments. This real-time feedback mechanism improves process reliability, maximizes resource recovery, increases product yields, and helps minimize operational expenses, contributing to both economic and environmental sustainability [1, 2].

3.2 Instruments for Waste Treatment

3.2.1 Physical Treatment Instruments

The physical treatment involves sorting and pretreatment. The sorting and pretreatment instruments play a crucial role in preparing materials for subsequent processing steps. In the context of waste management, these tools are used to separate various types of waste, ensuring that recyclable materials are properly sorted and nonrecyclables are directed to appropriate disposal methods. These instruments help in separating and preparing waste for further processing, recycling, or conversion.

3.2.1.1 Sorting Instruments

Sorting instruments help in screening and separating different materials from the waste with the help of various instruments. These materials are sorted on the basis of the physical and chemical properties of the waste components. The instruments separate the components as per their particle nature, size, shape, color, viscosity, and density. It includes conveyors screens, and trommel screens.

- **Conveyor Belt Systems:** Conveyor belt systems are used for transporting and sorting materials. It is a mechanical system used to transport materials and products from one location to another.
- **Trommel Screens:** A trommel screen, or rotary screen, is a mechanical device used for material separation, commonly employed in the mineral processing and solid waste management sectors. It features a rotating cylindrical drum with perforations, typically inclined at the input side to facilitate the movement and sorting of materials based on the size of waste. It is used extensively in solid waste management, composting plants, and biogas facilities. Trommel screens are ideal for handling heterogeneous and moist waste streams [3, 4].

3.2.1.2 Pretreatment Instruments

The instruments used for the pretreatment of waste help in reducing and resizing the waste. It helps in separating the different components of waste by physical pretreatment process. The pretreatment of waste is also done for medical and biological purposes by cleaning and sterilization processes. All these techniques and the instruments help to further break down the waste physically and convert it into much smaller particles.

3.2.1.2.1 Pretreatment for Separation of Solid from Solid

The equipment is used for the reduction of the size of waste material. Equipment such as mechanical conveyors, shredders, and screening systems are commonly used in this phase. Shredders and grinders can reduce the size of waste materials to make them easier to process. The shredders are machines that cut things, such as papers, documents, and other waste materials, into very small pieces. Grinder is a machine or tool that crushes the waste material or small shredded materials into much smaller shape, smooth particles, or materials through abrasion. It can also crush or grind waste material into smaller pieces, powder, or paste. These mechanical pretreatment instruments or separators include screens, conveyors, and optical sorters that use physical properties like size, shape, density, or color to separate materials (Figure 3.1).

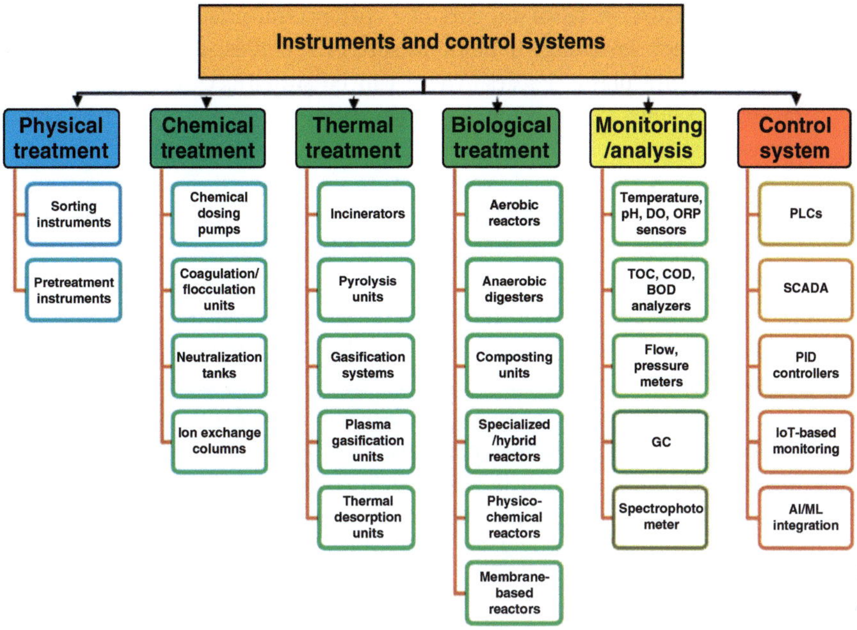

Figure 3.1 Flowchart of various instruments and control systems used for waste treatment.

Various other instruments that are used to sort and pretreat the waste material are as follows:

- **Waste Spiral Screen:** It separates materials based on size in landfill waste. It is designed for efficient three-way separation of solid waste. It operates by directing large-sized materials to move over the spiral shaft separator, while medium-sized waste follows the flow along the spiral shaft. Meanwhile, smaller particles fall through the gaps between the spiral blades, enabling effective classification based on size.
- **Chain Plate Feeder:** It is a strong impact-resistant instrument. It is suitable for a wide range of materials, such as mixed waste, residual waste, household waste, and landfill waste.
- **Magnetic Separators:** The design of the magnetic field in this system was selected based on the optimizing computerized simulation. It is used for recyclable metal recovery and helps in the removal of ferrous metals.
- **Waste Shredders/Fine Shredders:** It works on low speed, high torque, and strong shredding force. It works with multiple functions such as cutting, chopping, and shredding. It reduces particle size of waste and prepares organic and plastic waste for further treatment. It can easily perform the task of shredding mixed solid waste with complex components.
- **Grinders:** It is used for removing material from a workpiece via abrasion. It grinds any material in the form of powder or paste.

- **Optical Sorters:** It takes support of hyperspectral imaging system plus AI Machine Deep Learning. It helps to sort PET bottles, HDPE bottles, cans, cardboard, and plastic bottles. These can be sorted according to color and material. It helps to identify materials based on color, shape, or other optical characteristics.
- **Automated/X-ray Optical Sorters:** It uses sensors (IR, X-ray) to sort materials by type. It efficiently sorts inorganic inert materials in organic waste, including stones, bricks, metals, glass, and other impurities.
- **Eddy Current Separator:** It is a unique magnetic system design with strong anti-demagnetization ability. It helps in the automatic separation or removal of nonferrous metals and nonmetals such as aluminum and copper sorting.
- **Balers:** It works on the basis of hydraulic drive, high pressure, high efficiency, high stability, and efficient storage. It provides proper transport and converts waste into compact recyclable materials.
- **Metal Balers:** The system is designed to reduce metals based on hydraulic push, with options for manual operation or PLC-based automated control. It is suitable for baling scrap steel, scrap aluminum, iron, cans, stainless steel, and other metals.
- **AI Sorting Robot:** The sorting speed is fast and can achieve more than 5,500 suck-up sorting per minute. It has an ability to sort different recyclables simultaneously, such as plastic bottles, cardboard, cans, etc.

3.2.1.2.2 Pretreatment for Separation of Solid from Liquid

- **Sedimentation Tanks:** The sedimentation tank is also known as a clarifier. They help to separate solids from liquids due to the principle of gravity. The suspended solid at the bottom of the tank is separated from liquid by the decantation process. The liquid is decanted from the upper layer, whereas the separated solid is collected at the bottom of the tank and is known as sludge.
- **Centrifuges:** The centrifuges help in the separation of liquid from solid due to the principle of density difference and by using centrifugal force. A centrifuge is a device designed to apply centrifugal force to a sample, typically to separate its components based on density. This separation is accomplished by rapidly spinning the sample within a container, forcing heavier substances outward and allowing lighter ones to migrate toward the center. For instance, it can be used to isolate cream from milk or to separate solids from liquids. In laboratory settings, centrifuges often utilize sample tubes, where the spinning motion causes heavier particles to settle at the bottom, while lighter materials remain near the top. On a larger scale, industrial centrifuges are widely employed in manufacturing and waste treatment processes. They are particularly effective in settling suspended solids or separating liquids that do not mix, making them valuable tools for filtration and purification in various industries.
- **Filtration Units:** This technique involves the separation of solids from liquids by using filter membrane under applied pressure or vacuum. This filter membrane includes sand filters, membrane filters, and cartridge filters. There are

various instruments like rotary vacuum filter drum, horizontal filter press, vertical filter press, plate and frame filter press, and pressure leaf filter press. The instruments are used to remove solids from liquid waste material or effluent. The solid recovered is collected in the form of dry cake, which is further treated or utilized through biological processes.

3.2.1.2.3 Screening Units

This equipment helps to remove large debris from wastewater. Screening is one of the most fundamental and essential steps in solid waste and wastewater treatment processes. Screening equipment is used to remove large and coarse solids (e.g., plastics, rags, wood pieces, cans) from the waste stream before the primary treatment stage. Efficient screening protects downstream equipment from clogging, improves process performance, and enhances the longevity of the treatment system [2–4]. Screening systems can be broadly classified into manual, mechanical, and automated systems. These can be further categorized based on design and functionality:

- **Bar Screens:** It consists of parallel metal bars placed at an angle to the flow to trap large solids. These bar screens are of various sizes (>50 mm), medium (20–50 mm), and fine screens (<20 mm). It is applicable for preliminary treatment in municipal and industrial wastewater plants. The instruments are equipped with rakes and mechanical drives for cleaning. Bar screens reduce operational downtime and prevent blockages in pumps and pipelines.
- **Drum and Rotary Screens:** The drum and rotary screens are the cylindrical drum fitted with a mesh or perforated surface that rotates, allowing water to pass while retaining solids. It is applicable in fine screening of wastewater and biomass slurry. The continuous operation and automatic cleaning are their major advantages. Rotary screens are efficient for separating fibrous materials and work well in high-volume treatment systems.
- **Vibrating Screens:** It utilizes vibratory motion to segregate materials based on particle size. It is used in compost plants and recycling units for separating organic matter, plastic, glass, and inert materials. It efficiently classifies particle sizes, enhancing waste sorting and material recovery.
- **Spiral (Screw) Screens:** In spiral screens, waste moves along a rotating spiral shaft where different fractions are separated by size. It is applicable in municipal wastewater plants and industrial food waste separation. Its compact design and easy installation are major advantages of spiral screen. Screw screens combine conveying and screening, improving space and energy efficiency.
- **Dewatering Press:** The belt filter press is the instrument used for the dewatering process, and it reduces water content in sludge. A dewatering press is a mechanical system used to reduce the water content in sludge generated from wastewater treatment. By applying pressure, the device separates liquid from solids, producing a drier, more compact sludge cake that is easier to manage and dispose of. Different types of dewatering presses are available, including belt filter presses, screw presses, and plate-and-frame presses, each suited for specific treatment needs. These systems are widely used in municipal and

industrial wastewater facilities as well as in agricultural waste processing. Reducing the volume of sludge through dewatering helps to cut down transportation and disposal costs while improving the overall efficiency of downstream processes like composting, incineration, or landfill use. Additionally, it supports more sustainable waste handling by minimizing moisture-related environmental concerns.

- **Float Sink Tanks:** It separates materials based on density differences between materials in a liquid medium, typically water. The material is deposited onto the surface of the water in the tank. It is primarily used in plastic recycling and other material separation processes to separate materials based on their density [5, 6].

3.2.1.2.4 Drying and Dehydration Equipment

- **Dryers**
  It is used to remove moisture from organic waste, reducing volume and making it easier to handle. Drying plays a vital role in the treatment of Municipal Solid Waste (MSW). One of the primary goals of drying is to reduce the moisture content in the waste materials. High-moisture levels can significantly lower the calorific value (energy content) of the waste, making it less efficient as a fuel. By removing excess water, the energy density of the material increases, meaning more energy can be extracted per unit mass during combustion. Drying is a fundamental unit operation in the treatment of waste, particularly solid and organic fractions, as it facilitates improved handling, storage, and downstream processing. The reduction in moisture content significantly lowers the mass and volume of the waste, enhancing transport efficiency and reducing operational costs. Drying helps to minimize problems such as corrosion and harmful emissions that arise the problems during the burning of wet waste. Moisture in the waste can contribute to the formation of acidic gases or encourage the release of pollutants, both of which can damage equipment and harm the environment. By enhancing the physical and chemical properties of MSW and refuse-derived fuel (RDF), drying makes these materials more suitable for thermal treatment processes like incineration, coprocessing in cement kilns, or gasification. This step ultimately improves fuel efficiency, reduces environmental impacts, and supports the sustainable recovery of energy from waste. This continuous drying mechanism makes rotary drum dryers especially useful in processing bulk solids like fertilizers, minerals, biomass, and food products, where consistent drying and energy efficiency are critical. Rotary dryers are used primarily for chemical processing and mineral extraction; they are extraordinarily versatile machines. They are also widely used in agriculture, ceramics, dairy, food, lumber, paper, rubber, and other industries [7]. The dryers are the instruments used to reduce moisture content also in sludge or solid waste.
- **Rotary Drum Dryer**
  Rotary drum dryer is a type of industrial drying equipment used to reduce the moisture content of solid materials. It consists of a large, slightly inclined

cylindrical drum that slowly rotates around its horizontal axis. The drum is typically made of metal and is hollow inside. During operation, the wet material is introduced into the drum from the elevated end. As the drum rotates, the material gradually moves toward the lower discharge end due to gravity and the slight slope of the cylinder. Internally, the drum may be equipped with lifting flights or paddles that continuously lift and tumble the material, ensuring even exposure to the drying heat. Heat is supplied either directly or indirectly. In **indirect heating**, the drum is surrounded by a jacket or internal heating element through which hot water, steam, or another heating medium flows. This heat is transferred through the metal walls of the drum to the outer surface, where a thin layer of material is spread. The dryer uses a heating element to generate hot air, which is then circulated through the drum. Rotary drum dryers are low-temperature drying instruments that operate below 200 °C (depending on waste moisture). The air is heated to the desired temperature, which is usually around 150–180 °F (65.5–82.2 °C). Once the air is heated, the blower fan circulates it throughout the drum. The moisture evaporates as a result of this heat exchange, drying the material efficiently. It minimizes plastic thermal shrinkage and fire risks [8].

- **Wastewater Heat Recovery**
  This focuses on recovering heat from wastewater streams to preheat influent water or for other heating applications. It can significantly reduce energy consumption in wastewater treatment plants.

3.2.1.2.5 Pretreatment of Biological/Medical Waste

In biological and medical processes, pretreatment involves thorough cleaning and disinfection of instruments to eliminate biological contaminants before they undergo sterilization. This step is essential to ensure the safety and effectiveness of the sterilization process, preventing the spread of infections and maintaining hygiene standards (Table 3.1).

- **Manual Cleaning Tools**
  The manual cleaning is done using various tools such as brushes, sponges, and specialized cleaning equipment, and solutions are used to remove visible soil and debris.

Table 3.1 Biological instruments used for the pretreatment of waste.

S. No.	Tool name	Function
1	Rake	Removes large floating or settled solids from screens or tanks
2	Shovel	Lifts and removes settled solids or grit
3	Brush/broom	Cleans surfaces, screens, or equipment surfaces
4	Bucket	Collects and transports removed debris or sludge
5	Hand screen	Manually intercepts large solids from wastewater
6	Scraper	Removes accumulated sludge or scum from tank walls or floors

- **Ultrasonic Cleaners**
  Ultrasonic cleaners are specialized cleaning devices that utilize high-frequency sound waves in a liquid medium to eliminate dirt, grease, or contaminants from surfaces. While they are not used for treating wastewater directly, they play a vital role in maintaining laboratory instruments, probes, and precision tools involved in water analysis and treatment monitoring.
- **Washer-disinfectors**
  These are automated machines that clean and disinfect instruments. Washer-disinfectors are automated cleaning devices designed to clean and thermally disinfect reusable instruments, equipment, and containers, primarily in healthcare, laboratories, and food industries. They use a combination of high-pressure water jets, detergents, and heat to remove organic and inorganic contaminants, followed by a thermal disinfection cycle that kills microorganisms. These machines operate on programmed cycles to ensure consistent, validated, and hygienic cleaning standards. Although not typically used for bulk wastewater treatment, washer-disinfectors play a crucial role in the sanitization of tools used in water testing, clinical diagnostics, and environmental laboratories. Their use ensures the safe reuse of equipment, reduces the risk of cross-contamination, and supports compliance with health and safety regulations.
- **Autoclaves**
  It sterilizes medical waste using 121 °C for 15 minutes at high-pressure steam. This process involves steam sterilization. Mechanical Biological Treatment (MBT) paired with autoclave or converter systems "cooks" mixed waste with steam/heat under pressure (or atmospheric variable pressure) to sterilize and separate fiber, plastics, and metals, achieving volume reduction up to 60% and residual <10%.

3.2.2 Chemical Waste Treatment Instruments

There are certain chemicals that can be used to modify waste materials and improve their suitability for recycling or other processes.

3.2.2.1 Chemical Dosing Pumps

Chemical dosing pumps are precision devices used to inject accurate amounts of chemicals into a process system, typically in water and wastewater treatment, industrial processes, or laboratories. These pumps ensure controlled and consistent dosing of liquids such as acids, alkalis, disinfectants, or coagulants. They are designed to handle corrosive or hazardous fluids and can be manually or automatically operated, often integrating with sensors or control systems for precise flow regulation. Their reliability and accuracy help maintain optimal chemical concentrations, improve process efficiency, and ensure safety and compliance with environmental standards.

3.2.2.2 Coagulation and Flocculation Units

These instruments and the techniques treat the dissolved solids and remove them from the wastewater by adding acids or bases to the wastewater. The acid or base chemicals facilitate aggregation of small particles in the form of coagulants, and the process is known as coagulation. The chemicals, such as $FeCl_3$ and Alum, aggregate the coagulants and make big flocs. These flocs are comparatively bigger in size as compared to coagulants. These particles get settled down in the suspension and help to separate and clear the waste present in water or any waste effluent stream.

3.2.2.3 Neutralization Tanks

Neutralization tanks play a critical role in industrial wastewater treatment by regulating pH levels to ensure environmentally safe discharge. These systems are specifically designed to treat acidic or alkaline effluents, preventing corrosion of infrastructure and minimizing environmental impact. Through controlled chemical reactions, the tanks adjust the pH of wastewater to a near-neutral value around pH 7, using agents such as lime, sodium hydroxide, or calcium carbonate. The influent, whether acidic or basic, is directed into the tank where it interacts with the neutralizing substances. To resist the corrosive properties of the waste, these tanks are typically made from durable, chemical-resistant materials like polyethylene, polypropylene, or fiberglass-reinforced plastic (FRP). It is applicable for research laboratories, industrial plants, hospitals, pharmaceutical manufacturing, and wastewater treatment plants.

3.2.2.4 Ion Exchange Columns

Ion exchange is a promising method for treating wastewater by meeting discharge standards and enabling nutrient recovery in concentrated forms. Ion exchange columns are widely used in wastewater treatment to eliminate unwanted ions by replacing them with more desirable ones. This technique is particularly effective for removing contaminants such as heavy metals, radioactive substances, nitrates, and sulfates. The process involves passing wastewater through columns filled with specialized resins, either cation exchange resins (for positive ions) or anion exchange resins (for negative ions). As water flows through the resin bed, undesirable ions are captured, and the resin releases its own ions in exchange. Common applications include removing metals like lead, copper, and nickel from industrial effluents, extracting radioactive elements from nuclear waste, and softening water by eliminating calcium and magnesium. Additionally, ion exchange can recover valuable materials from wastewater, enhancing resource recovery. This method is highly selective, efficient, and regenerable, helping to reduce operational costs despite the high price of resins. As an example, researchers have assessed phosphorus removal using Purolite FerrIX A33E resin and nitrogen recovery with natural zeolite (clinoptilolite) [9].

3.2.3 Thermal Treatment Instruments

3.2.3.1 Incineration

This is a combustion process where waste is burned at high temperatures (850–1,000 °C) in the presence of oxygen. The incinerators help burn hazardous, medical, or municipal solid waste. Incineration drastically reduces waste volume and can destroy harmful substances. It's often used for energy recovery (generating electricity or heat) and for reducing the volume of waste going to landfills. The modern plants combust municipal or infectious waste and generate electricity or heat by reducing waste volume by approximately 95%.

3.2.3.2 Pyrolysis

Pyrolysis is a thermal decomposition process that occurs in the absence of oxygen. It breaks down waste materials into simpler compounds, producing gases, liquids, and a solid residue (char). The gaseous and liquid products can be further processed to recover energy or valuable materials. In this process, the pyrolysis units decompose organic materials at high temperatures in the absence of oxygen. Pyrolysis is a thermal decomposition unit that heats organic waste, such as plastics, in the absence of oxygen and inert atmosphere to produce bio-oil, biochar, and syngas.

3.2.3.3 Gasification

Gasification is similar to pyrolysis but involves a limited amount of oxygen. It converts solid waste into a combustible gas (syngas) that can be used for energy generation. Gasification is considered a cleaner alternative to incineration because it reduces the release of pollutants into the atmosphere. Gasification systems heat organic waste with a controlled amount of oxygen to produce syngas, which can be used for energy production.

3.2.3.4 Plasma Gasification

Plasma gasification uses plasma torch of high-temperature plasma to break down hazardous/municipal waste into syngas (H_2/CO), slag, metals, and minimal emissions. It can handle a wider range of waste materials, including hazardous waste, and produces a synthetic gas and a vitrified (glass-like) residue [10].

3.2.3.5 Thermal Desorption

This process uses heat to vaporize volatile organic compounds (VOCs) from waste materials, such as contaminated soil or sludge. The VOCs are then removed from the waste stream and can be further treated or recovered as a valuable compound [11].

Thermal treatment offers several advantages in waste management. The primary benefit is significant volume reduction, as the application of heat decreases the moisture content and enhances the degradability of waste, thereby minimizing the quantity that ultimately requires landfilling. Thermal processes such as incineration and gasification enable the recovery of energy, transforming waste materials into usable power. These methods are also effective in managing hazardous waste by breaking down or neutralizing harmful substances. By decreasing the overall waste volume, thermal treatment contributes to the conservation of landfill space. However, certain factors must be considered when employing these technologies. Environmental concerns, including the release of pollutants and the management of residual ash, must be carefully controlled. Moreover, the financial investment required for installation and operation can vary widely depending on the technology and scale of the facility. Lastly, the composition of the waste plays a crucial role in determining the efficiency and suitability of specific thermal treatment methods.

3.2.4 Biological Waste Treatment Instruments

Instruments used for organic waste valorization encompass a wide range of technologies designed to transform biodegradable waste into valuable end products like biogas, biofertilizers, and other bio-based materials. These tools can vary from basic composting setups to advanced systems such as anaerobic digesters (AD) and insect-based bioconversion units. Each of these instruments supports the sustainable management of organic waste by facilitating its conversion into useful resources, reducing environmental impact, and contributing to circular economy practices. The biological instruments are used for converting food, green, or biodegradable waste into compost, bioenergy, and bio-products.

3.2.4.1 Bioreactor/Fermenter

The bioreactors and the fermenters are used for microbial degradation of pollutants. They provide open or closed environments for microbial processes, where the pollutants are used as substrates and microbes catabolically break down the pollutants into simple compounds. The bioreactors, or the fermenters, are vessels with accessories such as agitators, impellers, spargers, baffles, sensors, ports, and control systems such as Proportional–Integral–Derivative (PID), Proportional–Derivative (PD), and Proportional–Integral (PI) controllers, and utility systems such as coolers, chillers, steam generators, and exchangers, as shown in Figure 3.2.

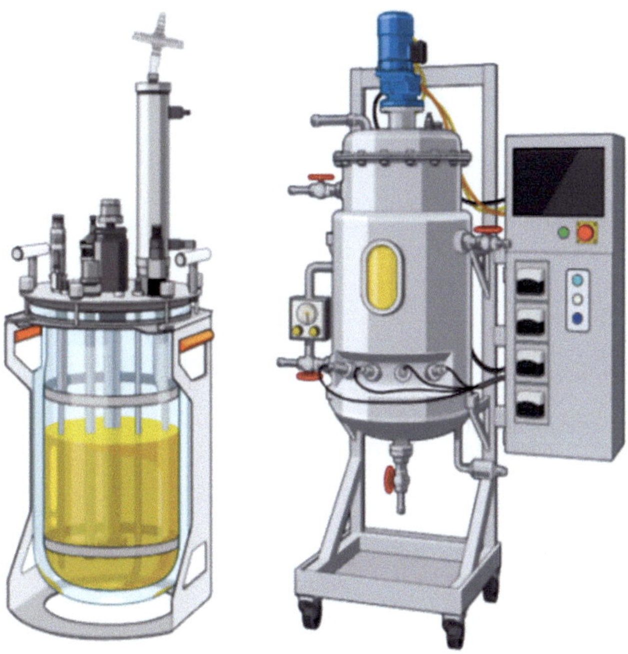

Figure 3.2 Bioreactors and its parts.

Different types of bioreactors applicable for various purposes are as follows:

3.2.4.1.1 Aerobic Biological Reactors

These utilize oxygen for microbial degradation of organic pollutants.

- **Continuous Stirred Tank Reactors (CSTRs)/Activated Sludge Systems**
  These reactors are homogeneous reactors with continuous input and output and are used in activated sludge processes. These activated sludge systems are commonly used aerobic biological reactors for domestic and industrial wastewater. These are conventional reactors containing suspended biomass, effective for treating low-strength wastewater (COD < 1000 mg/L) and capable of producing effluent with BOD levels below 30 mg/L [10]. The activated sludge process is a widely utilized biological method for the treatment of municipal and industrial wastewater. It is an aerobic biological treatment system designed to aerate wastewater in the presence of a mixed microbial culture and convert dissolved organic pollutants into biomass. The mixed culture, commonly referred to as biological flocs, primarily consists of bacteria and protozoa. This process plays a central role in secondary treatment, focusing on the biological degradation of organic matter and the removal of suspended solids, which can subsequently be separated from the treated effluent by gravity or filtration. The activated sludge system primarily comprises three essential units: the anaerobic equilibration, the aeration basin, and the sedimentation (clarifier) basin, so the treatment occurs in three primary stages (Figure 3.3).

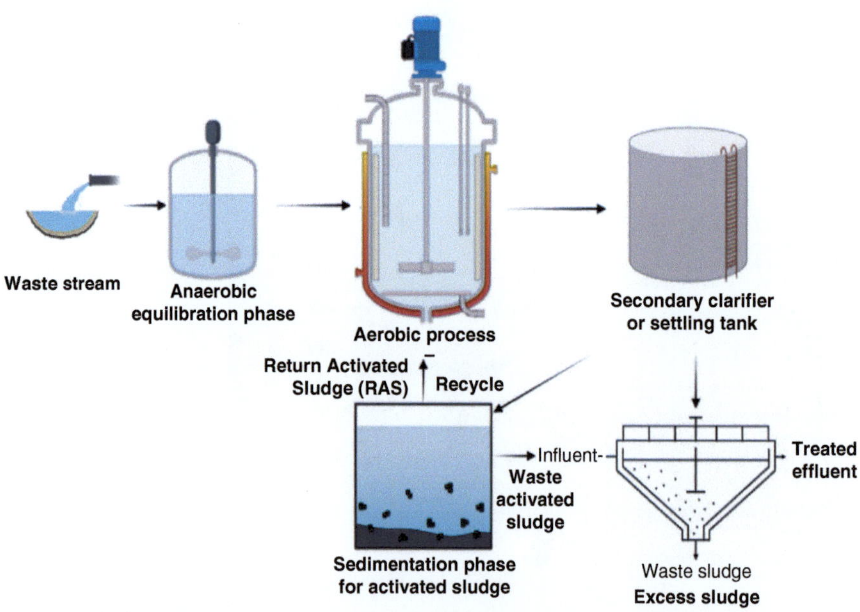

Figure 3.3 Diagrammatic flowchart illustrating activated sludge systems.

Stage I: It involves an anaerobic equilibration phase, where the incoming wastewater undergoes initial conditioning.

Stage II: In this stage, the mixture is transferred to an aeration basin where oxygen is supplied to facilitate microbial oxidation of organic compounds and ammonia. Within the aeration basin, wastewater is treated using a dense microbial community under aerated conditions. Retention times in kraft pulp mill facilities typically range between 15 and 48 hours, with modern systems often utilizing longer durations for enhanced treatment efficiency. Mechanical aeration systems are used to ensure proper mixing and oxygen delivery, with various aerator designs employed depending on site-specific requirements. During treatment, the wastewater is introduced into an aeration tank, where it is mixed with air or oxygen to facilitate microbial activity. The microorganisms metabolize the organic pollutants, converting them into more stable forms.

Stage III: After aeration, the mixture is directed to a secondary clarifier or settling tank for solid–liquid separation. It consists of a sedimentation phase, during which the biomass or the biological flocs (activated sludge) settle under gravity, separating the treated water from the activated sludge. A portion of the settled sludge is discharged as waste (excess sludge) and is known as return activated sludge (RAS), while the remainder is recycled back to the aeration tank to maintain a high concentration of active microbial population, while the excess, termed waste activated sludge (WAS), is removed for further processing and eventual disposal.

A frequent operational issue in activated sludge systems is sludge bulking, where the sludge loses its density and fails to settle effectively, leading to reduced effluent quality. Addressing sludge bulking requires continuous monitoring, experienced personnel, and in many cases, dedicated full-time staff to respond promptly. The system is susceptible to disturbances, generates a significant amount of biological waste sludge, and may suffer from operational instability without protective features such as an equalization basin. Additionally, it often involves high operating costs and requires ongoing system adjustments.

Therefore, it is essential to design treatment systems that offer several advantageous features, including high pollutant removal efficiency, effective process control, especially regarding oxygen usage, space efficiency, and relatively low capital investment. Alternative treatment technologies that are more compact and cost-effective than traditional activated sludge systems have been developed. Although operational experience with these alternatives is still limited, reports suggest that they can achieve pollutant reduction rates comparable to those of conventional systems [12–20].

Various configurations of the activated sludge system are employed, including package plants, oxidation ditches, deep shaft systems, jet aerators, surface-aerated basins, fine bubble aerators, submerged turbine aerators, and sequencing batch reactors (SBRs). The aeration itself can be achieved through several techniques, such as diffused air systems, mechanical surface aerators, or in some cases, pure oxygen injection. There are multiple configurations and design approaches available for implementing the activated sludge process. These variations differ in terms of the layout of the aeration basin, the type of aeration system employed, the structure of the clarifier, and the method of sludge recirculation. One notable variation is the pure oxygen activated sludge system, which utilizes pure oxygen or oxygen-enriched air in place of conventional atmospheric air to enhance treatment efficiency.

- **Composting System**
  Composting is a controlled aerobic process in which microorganisms break down organic solid waste in the presence of oxygen. These organic materials, typically rich in carbon, can include items such as food waste, grass clippings, fallen leaves, and garden or tree trimmings. It includes windrow composting and in-vessel or rotary composters (static-pile and aerated-pile). It is suitable for converting organic MSW into nutrient-rich compost.
  - Aerated static pile composting involves creating piles of organic waste and using forced aeration to facilitate decomposition.
  - In windrow composting, organic waste is piled in long rows (windrows) and periodically turned to ensure proper aeration and decomposition.
  - In-vessel composting is the enclosed system that controls temperature, moisture, and aeration for faster and more efficient composting.
- **Plug Flow Reactor (PFR)**
  The PFR is a rectangular, partially mixed system where solid or semisolid feed enters from one end and exits from the other, aided by gravity through inclined baffles. This design eliminates the need for mechanical mixing, resulting in low energy usage, a simple layout, and stable operation. However, it faces challenges such as sedimentation, which can reduce hydraulic and solid retention times, and poor thermal uniformity due to crust formation on the walls. It is primarily used for anaerobic fermentation of materials like sludge, vegetable waste, and cow dung.
- **Sequencing Batch Reactors (SBRs)**
  They are batch-operated activated-sludge reactors that combine filling, aeration, reaction, settling, and decantation phases in a single tank. They offer flexibility and space efficiency, ideal when effluent standards are strict.
- **Trickling Filter Reactor (TFR)**
  TFR uses a bed of media over which wastewater trickles and biofilm forms on media. A bioelectrochemical system (BES)-integrated TFR effectively treated wastewater, achieving over 98% removal of COD and total nitrogen at a 0.5 g-COD/L/day load and 8 L/h circulation. Coulombic efficiency reached up to 16%, and COD removal dropped without the BES, highlighting its contribution. The anodic biofilm supported diverse microbial communities, including Geobacter for electricity generation and others for organic and nitrogen degradation. This system offers efficient, low-sludge treatment for high-load wastewater.
- **Rotating Biological Contactors (RBCs)**
  In this reactor, the disks rotate through wastewater to support biofilm growth under aerobic conditions. The rotating disks provide aeration and surface for microbial growth. The advantages of RBCs include lower energy cost than

conventional aeration. The drawbacks include sensitivity to influent variability and the need for periodic sludge removal.

- **Moving Bed Biofilm Reactors (MBBRs)**
  These reactors use suspended plastic carriers to support microbial biofilm development. They can retrofit existing plants, maintain high biomass, reduce hydraulic retention time, and lower capital costs.
- **Aerated Lagoons**
  These are the open pond reactors with mechanical aeration. The complete-mix variants keep solids suspended (high energy), while partial-mix types allow some sludge settling (lower energy), trading off removal efficiency for cost-effectiveness [21–24].

3.2.4.1.2 Anaerobic Biological Reactors

Anaerobic Biological Reactors operate in absence of oxygen and convert organic matter to methane and CO_2.

- **Anaerobic Digesters (AD) Are CSTR-based Reactors**
  These are the early systems that are commonly used for the treatment of organic sludge, agricultural waste, and food waste to digest them at low rates with long retention times and large tank volumes. Anaerobic digestion is a biological process in which microorganisms decompose organic materials such as animal manure, food waste, and sewage sludge in an oxygen-free environment. This process occurs within an airtight container known as a digester or reactor, which is specifically engineered in different sizes and configurations to suit the type of feedstock and site conditions. Within these reactors, diverse microbes break down the organic matter, resulting in the formation of biogas and digestate, the solid and liquid by-products of the digestion process. A common practice within anaerobic digestion is co-digestion, where multiple types of organic feedstocks are processed together in the same reactor. These materials can include livestock manure, agricultural residues, food processing and household food waste, energy crops, and fats, oils, and greases (FOG) collected from restaurants (Figure 3.4).

  Co-digestion enhances biogas yield, especially when combining low-energy or hard-to-digest substances with high-energy organic inputs.
  - Digesters are the enclosed tanks where microorganisms break down organic matter in the absence of oxygen, producing biogas (primarily methane and carbon dioxide).
  - Biogas upgrading systems are the equipment used to purify biogas into biomethane, which is used as a renewable fuel. Biogas primarily consists of methane (CH_4) the same energy-rich compound found in natural gas that typically ranging from 50 to 75%, along with carbon dioxide (CO_2), hydrogen sulfide (H_2S), water vapor, and small traces of other gases. This methane-rich gas can be utilized similarly to conventional natural gas for producing heat, electricity, or even for powering cooling systems. Biogas can also be upgraded by removing noncombustible or less valuable components like CO_2, H_2S, and moisture, resulting in renewable natural gas. This purified gas can be

Figure 3.4 Anaerobic digester instrument showing digestion process.

injected into existing natural gas pipelines, compressed for use as a fuel in vehicles, or further refined to create alternative fuels, advanced energy carriers, or high-value biochemicals and bioproducts.

- Digestates are the nutrient-rich residue that remains after the anaerobic digestion process is complete. It consists of both solid and liquid fractions, which are typically separated and managed individually due to their distinct characteristics and potential uses. When properly processed, both components of digestates offer a range of valuable applications. The solid portion can be repurposed as animal bedding, used to produce compost, or applied directly to soil as an organic amendment or fertilizer. The liquid fraction is rich in nutrients and can also serve as a fertilizer. Additionally, digestate, particularly the solid component, can be used as a base material for creating bio-based products such as bioplastics. Utilizing digestate in these ways can generate additional income or reduce costs, enhancing the overall economic and environmental performance of anaerobic digestion and biogas projects [25, 26].

- **Anaerobic Filters (AFs)**
  These fixed bed reactors support anaerobic biofilm development on solid natural or synthetic media. They offer rapid startup and good solids retention, though long-term clogging remains a challenge.
- **Packed Bed Reactors (PBRs)**
  PBRs, also referred to as fixed bed reactors, are commonly utilized in catalytic processes. These reactors typically feature a cylindrical shell with convex ends and are most often arranged in a vertical orientation, enabling the reactants

to flow through the system by gravity. Within the reactor, a stationary bed of catalyst material is held in place, allowing reactions to occur as the reactants pass through. These systems operate under heterogeneous conditions, where the catalyst and reactants exist in different phases [27].

- **Anaerobic Baffled Reactor (ABR)**
  An ABR is a wastewater treatment system designed with a series of baffles that direct the flow of wastewater upward through several compartments within the tank. This configuration improves the contact between the wastewater and anaerobic microorganisms, resulting in more effective digestion and removal of organic pollutants than conventional septic tanks. ABRs are recognized for their durability, capacity to withstand fluctuations in load, and adaptability to treating different kinds of wastewater, such as domestic sewage, greywater, and certain industrial effluents. It has compartments that separate acidogenic and methanogenic phases.
- **Upflow Anaerobic Sludge Blanket (UASB)**
  In these reactors, the wastewater flows in upward direction through a dense blanket of granular sludge. These are efficient COD removal with compact footprint. The biogas production is the major benefit of UASB [28, 29].
- **Expanded Granular Sludge Bed (EGSB)**
  It is an enhanced version of UASB, allowing higher organic loadings and shorter hydraulic retention times, while retaining granular biomass. It is used with higher upflow velocity and efficiency.

3.2.4.1.3 Physicochemical Reactors

- **Ion Exchange Column (IEC):** It removes heavy metals, ammonia, and other ions using ion-exchange resins [7].
- **Advanced Oxidation Reactor (AOR):** It uses ozone, hydrogen peroxide, UV, or Fenton's reagent to oxidize contaminants.
- **Electrocoagulation Reactor (ECR):** It applies electric current to induce coagulation of suspended solids.
- **Chemical Dosing Reactor (CDR):** It injects chemicals such as alum and lime to facilitate precipitation or neutralization.
- **Neutralization Reactor:** It adjusts pH by chemical dosing in industrial wastewater [11, 30–32].

3.2.4.1.4 Membrane-based Reactors

These reactors combine biological treatment with membrane filtration. There are various reactors, such as Membrane Bioreactors (MBRs) and Membrane Anaerobic Bioreactor (AnMBR).

- **Membrane Bioreactors (MBRs):** The MBR enables high MLSS (Mixed Liquor Suspended Solids) levels and compact footprint for aerobic systems. It integrates activated sludge with membrane filtration for solid–liquid separation.
- **Membrane Anaerobic Bioreactor (AnMBR):** This is anaerobic version of MBR; it is applicable in recovering and capturing methane and produces less sludge of low volume [33–35].

3.2.4.1.5 Emerging/Specialized Reactors

It uses chemical or physical processes rather than biological degradation. It has advanced designs for specific applications.

- **Fluidized Bed Reactors (FBRs)**
  The microbial cells or enzymes are immobilized on very lightweight solid particles. When these solid particles are inoculated in waste media and the air is supplied from the bottom, these solid particles along with the fluid flow in upward direction. The media suspended in fluidized motion improves contact and degradation rate of the waste effluent. This process enhances mass transfer and temperature control and is applied in certain wastewater or chemical treatment applications.
- **Photocatalytic Reactors (PWTRs)**
  This reactor uses UV or visible-light-activated catalysts (e.g., TiO_2) to oxidize organic contaminants. The slurry modes offer high reactivity but tough separation, while immobilized types are easier to recycle but have lower contact efficiency. The reactor is used for algae and photosynthetic organisms for nutrient recovery or CO_2 fixation [36].
- **Electro-oxidation Reactors (EORs)**
  This reactor removes pollutants through anodic oxidation. The flow-through electrode designs significantly improve mass transfer and energy efficiency compared to flow-by models [37].
- **Thermal–Chemical/Plasma Gasification**
  In this process, high-temperature plasma converts municipal or industrial waste into syngas and slag. It is applied to hazardous, biomass, or refuse-derived waste streams.
- **Hybrid/Integrated Systems**
  The hybrid anaerobic–aerobic reactor systems integrate anaerobic and aerobic zones (e.g., in wetlands or sequential reactors) and enable both energy recovery (via biogas) and improved nutrient removal. It is useful especially for landfill leachate and high-strength wastewater (Table 3.2) [38, 39].

Table 3.2 List of various categories of reactors and their functions.

S. No.	Category	Reactor types	Functions
1	Aerobic	CSTR (activated sludge bioreactor), PFR, TFR, SBR, RBC, MBBR, aerated lagoons	Efficient BOD/COD removal, flexible configurations, space/energy tradeoffs
2	Anaerobic	CSTR, AD, AFs/PBR, ABR, UASB, EGSB	High COD loading, biogas generation, lower sludge production
3	Physicochemical	IEC, AORs, ECRs, CDRs, neutralization reactor	It removes chemicals and metals by adding chemicals to change the pH of the suspension and precipitate or neutralize the waste

S. No.	Category	Reactor types	Functions
4	Membrane-based	MBR, AnMBR	High effluent quality, compact, methane recovery (AnMBR)
5	Emerging/ specialized reactors	FBRs, PWTRs, ECRs/EORs, thermal–chemical/plasma gasification, hybrid anaerobic–aerobic integrations	Target emerging or recalcitrant pollutants, higher energy or capital investment, combined advantages: energy recovery + nutrient removal

3.3 Monitoring and Analyzing Instruments

There are various critical parameters that need to be monitored during the fermentation or digestion process. It is also required to measure these parameters so that they can be controlled to conduct the efficient process. These instrumentations provide real-time monitoring and data acquisition of CPPs. The details of these instruments and their functions are mentioned in Table 3.3.

3.3.1 Materials Recovery Facilities

Materials recovery facilities (MRFs) are designed to sort and separate different types of recyclable materials, including some organic waste. These instruments and technologies play a vital role in converting organic waste into valuable resources, contributing to a more sustainable and circular economy. The infrastructure involves combining sorting lines, shredding, baling, or pelletizing to produce RDF and recover recyclable fractions (Table 3.4).

3.3.2 Energy Recovery and Waste-to-energy Instruments

An integrated waste-to-energy network is formed through collaboration among sectors that generate and utilize energy from waste. This system combines incineration, methane fermentation, and feedstock recycling to enhance energy recovery efficiency. Modeling revealed that the most effective configuration involved using methane to superheat boiler steam along with the application of refuse-derived paper and plastic fuel (RPF). The study also highlights key technological and policy considerations crucial for advancing low-carbon urban transitions. There are some instruments that are used in systems for converting waste to electricity, heat, or fuel (Table 3.5) [40, 41].

3.3.3 Instruments in Water and Leachate Treatment

The various instruments are used during water and leachate treatment. The water and leachate purification monitoring and analysis are essential for sustainable zero-waste operations, especially in landfills and compost sites (Table 3.6).

Table 3.3 List of instruments used for analyzing the critical parameters.

S. No.	Parameter	Instrument used	Purpose
1	Temperature	Thermocouples, Resistance Temperature Detectors (RTDs)	Monitor and ensure optimal conditions for microbial/enzymatic activity
2	pH	pH probes	Monitor and measure acid/base conditions for microbial viability and waste materials
3	DO (Dissolved Oxygen)	DO meters/DO sensors	Essential for monitoring aerobic biological treatment and aerobic degradation
4	Conductivity	Conductivity meter	Assesses ionic content in wastewater
5	Redox potential	Oxidation-Reduction Potential (ORP) meters	Evaluate oxidation/reduction balance in AD
6	Biogas composition	Gas chromatographs, IR sensors	Analyze methane and CO_2 levels for process efficiency
7	Flow rate	Ultrasonic or magnetic flow meters	Measure substrate input/output
8	Pressure	Pressure transducers	Monitor gas production systems
9	Turbidity/optical density	Turbidity meter/spectrophotometers	Measures the cloudiness of water to estimate suspended solids. Monitor biomass growth and reactor turbidity
10	Total organic carbon	TOC analyzer	Detects the amount of organic carbon in water or waste samples
11	Biochemical oxygen demand	BOD analyzer	Determines organic pollution load
12	Chemical oxygen demand	COD analyzer	Measures total quantity of chemicals in water that can be oxidized

Table 3.4 List of instruments for monitoring materials recovery facilities (MRFs).

S. No.	Instrument	Purpose	Application
1	Compost maturity analyzer	Measures CO_2, temperature, and moisture	Determines readiness of compost
2	pH and temperature probes	Measure pH and temperature	Composting, anaerobic digestion control
3	Gas chromatograph (GC)	Detects biogas composition	Methane and CO_2 analysis in digesters
4	Bioreactors with sensors	Controlled microbial degradation	Production of bioplastics, enzymes, ethanol
5	Respirometer	Measures biological activity	Efficiency of biodegradation

Table 3.5 List of instruments for monitoring during waste-to-energy recovery.

S. No.	Instrument	Purpose	Application
1	Calorimeter	Measures calorific value of waste	Assess energy potential of dry waste
2	Flue gas analyzer	Measures emissions from incineration	Air pollution control
3	Thermocouples and IR sensors	Temperature monitoring	Incinerators, pyrolysis units
4	Syngas analyzer	Checks CO, H_2, CH_4 content	Gasification systems for energy recovery
5	Microbial fuel cell (MFC) sensors	Monitor bioelectricity generation	Bioelectrochemical conversion of waste

Table 3.6 List of instruments for monitoring water and leachate treatment.

S. No.	Instrument	Purpose	Application
1	Turbidity meter	Measures clarity	Effluent monitoring
2	TOC (total organic carbon) analyzer	Measures organic load	Wastewater purification quality
3	Ammonia and nitrate sensors	Nutrient tracking	Leachate treatment

3.3.4 Smart Zero-waste Management Tools

The Think Zero Waste (TZE) app is a smart mobile solution that connects users with recycling companies and paper waste markets for efficient waste management. It allows users to schedule pickups, track collections in real time, and earn rewards for recycling. Using AI and ML, the app classifies waste through image recognition and routes it to appropriate recyclers or authorities. It also features predictive analytics, optimized collection routes, and personalized profiles. Through gamification and real-time GPS tracking, TZE supports local recyclers and municipal coordination, promoting sustainable practices and encouraging community participation in a circular economy. It is used in modern, automated, and Internet of Things (IoT) enabled systems to reduce and manage waste generation (Table 3.7) [42].

3.4 Control Systems in Valorization

Control systems are essential in valorization processes, which aim to extract maximum value from resources, including waste. They optimize operations by employing feedback loops, adaptive strategies, and robust communication networks. These

Table 3.7 List of instruments used as a smart zero-waste management tools.

S. No.	Instrument/tool	Function	Use case
1	IoT waste bins with fill sensors	Monitor bin levels and send alerts	Optimize collection routes
2	RFID trackers	Track recyclable materials	Product lifecycle and circular economy
3	SCADA systems	Centralized monitoring and control	Real-time data from waste treatment units
4	GIS & drones	Spatial data collection	Mapping waste zones, illegal dumpsites

systems enhance efficiency, ensure better resource utilization, and support sustainable waste-to-value conversion. By continuously monitoring and adjusting process parameters, control systems significantly improve the effectiveness of valorization efforts. These systems automate and regulate the process to maintain ideal conditions during the valorization of waste material:

- **PID Controllers**
  Proportional–Integral–Derivative (PID) controllers are widely used control mechanisms in engineering systems to regulate variables such as temperature, pressure, speed, and flow. These controllers operate by continuously calculating the error between a desired setpoint and a measured process variable, then applying a correction based on proportional, integral, and derivative terms.
  - Proportional (P) control provides an output value that is directly proportional to the current error. The larger the error, the stronger the correction applied, which helps in reducing the response time.
  - Integral (I) control addresses the accumulated past errors by integrating the error over time. This ensures that the controller can eliminate steady-state errors that the proportional term alone cannot resolve.
  - Derivative (D) control predicts future errors by considering the rate of change of the error. It acts as a damping function and helps to reduce overshooting and oscillations, improving overall system stability.

The combination of these three components allows PID controllers to achieve accurate and stable control in dynamic systems. They are simple to implement and widely applicable, making them a cornerstone in industrial automation, robotics, chemical processing, and environmental systems. There are different types of control systems such as feedback, feed forward, and cascade control systems. Modern PID controllers may include features such as auto-tuning, adaptive control, and digital implementation, allowing them to handle complex and nonlinear processes with higher efficiency.

- **Programmable Logic Controllers**
  A programmable logic controller (PLC) is a dedicated digital computer commonly used in industrial automation to manage and control machinery and

production processes. It collects input signals from sensors, processes the information using predefined logic, and then sends output signals to operate devices such as motors, valves, and actuators. PLCs are built to withstand tough industrial conditions and provide greater flexibility, reliability, and ease of programming compared to conventional relay-based control systems. It executes control logic to maintain setpoints for temperature, pH, etc.

- **Supervisory Control and Data Acquisition**
  Supervisory Control and Data Acquisition (SCADA) refers to a system designed for monitoring and managing industrial processes, particularly those spread across large areas. It integrates computers, communication networks, and graphical user interfaces (GUIs) to facilitate real-time control and data collection. SCADA systems play a vital role in numerous industries by allowing centralized oversight and management of operations that are distributed across different locations. SCADA systems are generally classified into four major types: Monolithic, Distributed, Networked, and IoT-integrated. These categories represent the progressive development of SCADA technology, influenced by improvements in computing, communication networks, and automation tools. Each system type incorporates components such as sensors, actuators, and human-machine interfaces (HMIs) to enable real-time monitoring and control of industrial processes.
- **IoT-based Monitoring Systems**
  IoT-based monitoring systems are advanced solutions that utilize interconnected sensors, devices, and networks to collect, transmit, and analyze real-time data from various environments. These systems are designed to enhance visibility, control, and decision-making across a wide range of sectors, including agriculture, healthcare, manufacturing, energy, and environmental management. At the core of an IoT monitoring system are smart sensors and actuators that continuously track parameters such as temperature, pH, agitation, dissolved oxygen, pressure, humidity, or machinery performance. These devices communicate wirelessly through protocols like Wi-Fi, LoRa, Zigbee, or cellular networks, transmitting data to a central platform, typically cloud-based, for processing and storage. The advantage of IoT-based monitoring is its ability to provide real-time alerts and insights, enabling users to detect anomalies, prevent failures, and optimize operations. In agriculture, IoT sensors can monitor soil moisture and weather conditions, while in industrial settings, they can detect equipment malfunctions or deviations from normal operating conditions. These systems often incorporate data analytics, artificial intelligence, and machine learning to interpret complex data patterns and support predictive maintenance and automation. Remote accessibility through mobile apps or web dashboards further enhances user convenience and operational efficiency. IoT-based monitoring systems offer a transformative approach to data-driven management, ensuring improved accuracy, reduced operational costs, and enhanced system performance across various domains.
- **AI/ML Integration**
  The integration of AI and ML in waste treatment control systems enhances automation, real-time monitoring, and adaptive decision-making. These

technologies analyze complex datasets such as pH, BOD, COD, and biogas levels to optimize operational parameters and predict system failures. AI-driven models help improve energy efficiency, reduce chemical usage, and maintain regulatory compliance. In biological treatment systems, machine learning enables dynamic control of aeration, dosing, and retention times. Advanced systems also support digital twins for simulation and forecasting. Overall, AI/ML enhances efficiency, sustainability, and reliability in modern waste management. It is a machine learning algorithm analyzing historical data to optimize microbial performance and reduce downtime [43, 44].

3.5 Modern Screening Equipment

Modern screening equipment in waste treatment employs advanced technologies to separate debris of varying sizes from wastewater and solid waste streams. Common types include mechanical bar screens, fine screens such as drum and step screens, and microscreens for targeted applications. These systems improve overall treatment efficiency by minimizing manual handling and preventing damage to downstream machinery. Their use ensures smoother operations and extends the lifespan of subsequent treatment components. The modern screening equipment can be integrated with the following systems:

- Automated cleaning systems
- Level sensors to control flow rates
- Wear sensors to detect screen damage
- Remote monitoring systems for process control
- SCADA systems for process control

Such integration improves operational efficiency, reduces manual labor, and minimizes equipment downtime [45].

3.6 Challenges and Future Trends

- Biofouling of sensors
- Lag in real-time microbial activity detection
- Need for miniaturized, multiparameter biosensors
- Integration of digital twins and AI-based predictive controls for fully autonomous waste valorization plants

3.7 Conclusion

A true zero-waste future can only be realized through the seamless integration of diverse technologies, empowered by precise and reliable instrumentation. From initial waste segregation to the final stage of valorization, advanced tools such as gas analyzers, spectrophotometers, bioreactor sensors, and smart IoT systems enable accurate monitoring, control, and optimization. These instruments ensure that every waste stream—whether organic, inorganic, electronic, or chemical—is efficiently recycled, reused, or transformed into energy and high-value products. By enhancing process efficiency, ensuring environmental compliance, and maximizing resource recovery, they transform waste from a liability into a valuable asset. This technology-driven approach not only mitigates emissions and environmental impacts but also reinforces the foundations of a circular economy, where resources are continuously cycled back into productive use. In this way, advanced instrumentation serves as the backbone of sustainable waste management, turning the vision of zero waste from an aspiration into an achievable reality.

References

1 Pérez-Parra, J.I., González-Martínez, S., and Gonzalez, F.J. (2023). Biological wastewater treatment technologies: a comprehensive review. *Environ Sci Pollut Res.* 30 (14): 19325–19340.

2 Tchobanoglous, G., Burton, F.L., and Stensel, H.D. (2014). *Wastewater Engineering: Treatment and Reuse*, 5e. New York: McGraw-Hill Education.

3 Spellman, F.R. (2013). *Handbook of Water and Wastewater Treatment Plant Operations*, 3e. Boca Raton: CRC Press.

4 Chen, Y.S., Hsiau, S.S., Lee, H.Y., et al. (2010). Size separation of particulates in a trommel screen system. *Chem Eng Process.* 49 (11): 1214–1221.

5 Metcalf & Eddy, AE. (2014). *Wastewater Engineering: Treatment and Resource Recovery*. New York: McGraw-Hill Education.

6 Metcalf & Eddy, Inc (2003). *Wastewater Engineering: Treatment and Reuse*, 4e. New York: McGraw-Hill.

7 Singh, R., Bhunia, P., and Dey, A. (2021). A comprehensive review on operation and performance of biofilters for municipal wastewater treatment. *J Environ Chem Eng.* 9 (4): 105379.

8 Smith, R., Brown, A., and Nguyen, T. (2020). Thermal drying of municipal solid waste and its impact on fuel quality and emissions. *Waste Manag.* 102: 409–417.

9 Ruiz-Cosgaya, L., Izquierdo, W.A., Martínez-Guijarro, R., et al. (2024). Ion exchange columns: a promising technology for nitrogen and phosphorus recovery. *J Environ Manage.* 370: 122719.

10 Gerrity, D., Stanford, B.D., Trenholm, R.A., and Snyder, S.A. (2011). An evaluation of a pilot-scale nonthermal plasma advanced oxidation process for trace organic contaminant degradation. *Water Res.* 45 (13): 4227–4237.

11 Pere, J., Siika-aho, M., Buchert, J., and Viikari, L. (1993). Effects of a thermochemical treatment on the enzymatic degradation of lignocellulose. *Bioresour Technol.* 45 (3): 193–198.

12 Bajpai, P. (2017). External treatment technologies used for pulp and paper mill effluents. In: *Pulp and Paper Industry* (ed. P. Bajpai), 71–92. Elsevier.

13 Thompson, G., Swain, J., Kay, M., and Forster, C.F. (2001). The treatment of pulp and paper mill effluent: a review. *Bioresour Technol.* 77 (3): 275–286.
14 Diez, M.C., Castillo, G., Aguilar, L., and Mora, M.L. (2002). Operational factors affecting removal of organochlorinated compounds in activated sludge systems. *Environ Technol.* 23 (3): 275–284.
15 Mahmood, T. and Elliott, A. (2006). A review of secondary activated sludge systems: performance evaluation and emerging technologies. *J Environ Manage.* 78 (2): 207–215.
16 Nakamura, M., Takeda, K., Fukami, K., and Sakurai, Y. (1997). Evaluation of activated sludge process for treatment of wastewater containing lignin derivatives. *Water Sci Technol.* 35 (2–3): 197–204.
17 Kostamo, A. and Kukkonen, J.V.K. (2003). Removal of resin acids and sterols from pulp mill effluents by activated sludge treatment. *Water Res.* 37 (13): 2813–2820.
18 Elliott, A. and Mahmood, T. (2012). Evaluation of activated sludge systems treating pulp and paper mill effluent. *Water Res.* 46 (3): 795–804.
19 Kaluža, L., Heath, C., and Brečko, D. (2014). Biological wastewater treatment in the paper industry. *Environ Technol.* 35 (19): 2436–2445.
20 Sarlin, E., Lehtinen, K.J., and Mattsson, K. (1999). Toxicity reduction of pulp and paper mill effluents in activated sludge treatment. *Water Sci Technol.* 40 (11–12): 45–52.
21 Nguyen, L.N., Hai, F.I., Price, W.E., et al. (2013). Removal of emerging contaminants by activated sludge processes. *Bioresour Technol.* 129: 640–648.
22 Fernandes, J., Ramisio, P.J., and Puga, H. (2022). *A Comprehensive Review on Various Phases of Wastewater Technologies: Trends and Future Perspectives.* MDPI.
23 Zhao, Y. and Ran, W. (2021). Anaerobic fermentation process for biomethane production from vegetable waste. In: *Biomethane Production from Vegetable and Water Hyacinth Waste*, 1e (ed. Y. Zhao and W. Ran), 1–62. Elsevier.
24 Liang, Q., Yamashita, T., Koike, K., et al. (2020). A bioelectrochemical-system-based trickling filter reactor for wastewater treatment. *Bioresour Technol.* 315: 123798.
25 Rajagopal, R., Massé, D.I., and Singh, G. (2013). A critical review on inhibition of anaerobic digestion process by excess ammonia. *Bioresour Technol.* 143: 632–641.
26 He, Y., Xu, P., Li, C., and Zhang, B. (2018). High-rate anaerobic reactors: evolution and future prospects. *Water Res.* 136: 118–133.
27 Levenspiel, O. (1999). *Chemical Reaction Engineering*, 3e. New York: Wiley.
28 Agridiotis, V., Forster, C.F., and Carliell-Marquet, C. (2007). Treatment of a primary sludge and industrial wastewater mixture in an UASB reactor. *Bioresour Technol.* 98 (1): 122–127.
29 Lettinga, G., van Velsen, A.F.M., Hobma, S.W. et al. (1980). Use of the upflow sludge blanket (USB) reactor concept for biological wastewater treatment. *Biotechnol Bioeng.* 22 (4): 699–734.
30 Scott, C.D. and Ollis, D.F. (1995). Integration of chemical and biological processes for wastewater treatment. *Environ Prog.* 14 (2): 88–96.
31 Novak, J.T., Chon, D.H., Curtis, B.A., and Doyle, M. (1998). The role of iron and aluminium in biosolids conditioning. *Water Environ Res.* 70 (1): 31–36.

32 Parker, D.S. (2016). Introduction to advanced oxidation processes (AOPs) for wastewater treatment. *Water Environ Res.* 88 (10): 1766–1785.

33 Wang, Z., Ma, J., and Wang, Z. (2022). Advances in membrane bioreactor systems for wastewater reuse: a critical review. *J Membr Sci.* 658: 120758.

34 Show, K.Y., Lee, D.J., and Tay, J.H. (2013). Membrane filtration for wastewater treatment. *J Chem Technol Biotechnol.* 88 (2): 267–276.

35 Judd, S. (2010). *The MBR Book: Principles and Applications of Membrane Bioreactors for Water and Wastewater Treatment*, 2e. Oxford: Elsevier.

36 Danfá, S., Martins, R.C., Quina, M.J., and Gomes J. (2021). Supported TiO_2 in ceramic materials for the photocatalytic degradation of contaminants of emerging concern in liquid effluents: a review. *Molecules* 26 (17): 5363.

37 Zhao, J., Wang, X., and Xu, J. (2019). Electrochemical treatment of wastewater: a review on reactor design. *Water* 11 (5): 1048.

38 Mishra, S., Singh, V., Ormeci, B., et al. (2023). Anaerobic-aerobic treatment of wastewater and leachate: a review of process integration, system design, performance and associated energy revenue. *J Environ Manage.* 327: 116898.

39 Feng, Q., Zhang, X., and Xu, G. (2023). Hybrid anaerobic-aerobic processes for high-strength wastewater treatment. *Environ Technol Rev.* 12 (1): 69–89.

40 Arena, U. and Di Gregorio, F. (2021). Waste-to-energy through gasification: a review of current status and emerging trends. *Renew Sustain Energ Rev.* 139: 110700.

41 Ohnishi, S., Fujii, M., Ohata, M. (2018). Efficient energy recovery through a combination of waste-to-energy systems for a low-carbon city. *Resour Conserv Recycl.* 128: 394–405.

42 Sumithra, M., Brindha, A., Archana, P., and Deepika, K. (2025). Smart waste management with AI powered classification and routing: the Think Zero Waste app (TZE APP). *2025 3rd International Conference on Communication, Security, and Artificial Intelligence (ICCSAI)* (4 April 2025, Vol. 3, pp. 1–5). IEEE.

43 Karthikeyan, O.P., Venkata Mohan, S., Kumar, G. (2020). Real-time monitoring and AI-based control for smart bioprocessing in wastewater systems. *Bioresour Technol.* 123476. 54 (1): 23–31.

44 Boe, K.(2011). Instrumentation and control of anaerobic digestion processes. *Water Sci Technol.* 63 (5): 1037–1044.

45 Musarat, M.A., Khan, A.M., Alaloul, W.S., et al. (2024). Automated monitoring innovations for efficient and safe construction practices. *Results Eng.* 22: 102057.

Part II

Processes and Techniques for Valorization of Different Wastes and Achieving Zero Waste

4

Bioprocess Techniques: A Profitable Industrial Approaches to Achieve Zero Waste

4.1 Introduction

"Zero waste" term was first applied in 1973 by Dr. Paul Palmer; he used this term to recover chemicals from waste [1]. Zero waste is a concept that is widely studied in multidiscipline. In waste management research, the zero-waste approach is the most controversial issue recently [2, 3]. Zero waste (ZW) is defined as "a system to design and management of products and processes for waste removal and material retention and rehabilitation, not burning or burying." This definition was given by Zero Waste International Alliance (ZWIA) in 2004. The zero-waste concepts are the utilization of waste via a sustainable biological or chemical process design for the conservation of resources and consumption practices for the recovery of maximum waste material to minimize pollution [4]. It does not include waste treatment and disposal and management such as landfill, incineration, burning, or burying. It is least feasible that all the waste will be consumed and there will be no leftover, and nil incineration and nil landfill targets can be achieved. To achieve zero-waste objectives, the industrial approach can be applied with the 3R concept (reduction, recycling, and reuse) to consume and utilize the resources economically with minimum environmental degradation. A zero-waste approach will considerably decrease the quantity of waste sent to landfills, which will reduce pollution and greenhouse gas emissions. Achieving successful zero-waste targets is required to establish an effective planning process. The chemical techniques are well in focus but alongside the bioprocess techniques are also the developing approaches for the utilization of waste to achieve zero waste as depicted in Figure 4.1. The biological techniques are also environmentally friendly.

- Bioprocess broadly includes fermentation, bioremediation, biological degradation, bioleaching, and biosorption of waste.
- Bioprocess involves bioreactors and downstream processing instruments for the process to achieve zero waste.

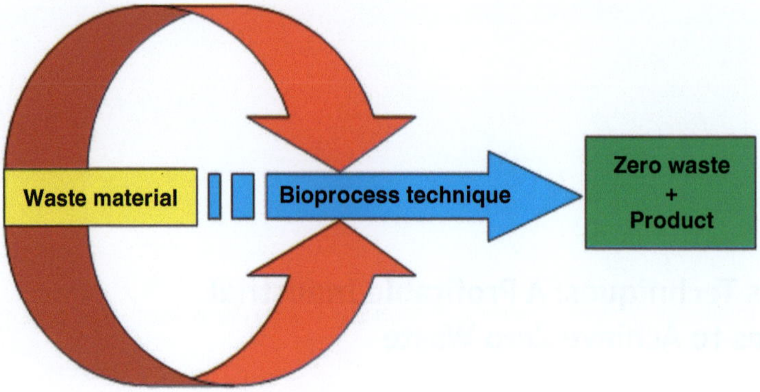

Figure 4.1 Process to achieve zero waste.

- These emerging technologies result in zero waste and also an economic benefit.
- It is required to promote and commercialize the research and application of zero-waste biotechnologies.

4.2 Bioprocess Approach for Zero Waste

Biotechnology is an emerging waste management technique to achieve zero waste. The role of biotechnology reduces waste and creates a zero-waste environment. Bioprocess technologies enhance the advances in zero-waste strategies. To achieve zero-waste, biotechnology effectively provides the path to creating sustainable products from waste. There are various bioprocess approaches for utilizing waste to achieve the zero-waste target. The bioprocess technologies include anaerobic co-digestion, biorefineries, biofuel production, distillation, extraction, fermentation, immobilized enzymes, integrated biosystems, membrane bioreactors, microbial fuel cell technology, nanobiomaterials, and various others. These bioprocess technologies utilize waste as a substrate that is converted into a product while using microorganisms such as bacteria, fungi, yeast, and algae/microalgae as shown in Figure 4.2. These bioprocess techniques are counted as a profitable approach for zero-waste management. Bioprocess is an advanced perspective for zero waste.

4.3 Waste Material Utilized by Bioprocess Techniques

The zero-waste approach can utilize sustainable materials such as sugary materials, lignocelluloses, and municipal waste. This biomass can be converted into valuable products such as animal feed, composites, bio-alcohol, biofuels, enzymes, fine chemical, perfumes, pieces of stuff, and pulp and paper. The details of the waste material are listed in Table 4.1. To make the bioprocess economically feasible and profitable and to increase reasonable industrial implementation to utilize zero waste, overall cost optimization is required for the advancement of bioeconomy [5].

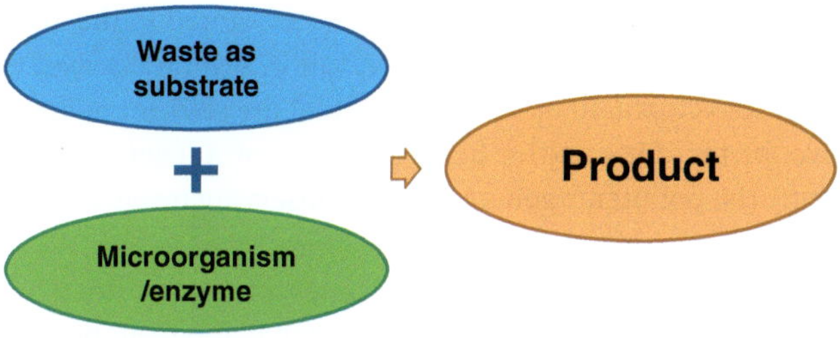

Figure 4.2 Diagrammatic representation of bioprocess technique for utilizing waste to achieve the zero waste.

Table 4.1 List of waste materials for utilization to achieve the zero waste.

S. No.	Waste material category	Waste material	Product produced
1	Sugary material	• Bagasse • Bagasse fly ash • Molasses • Press mud • Sugarcane trash • Sugar beet mud • Sugar beet pulp • Whey	Bio-alcohol, biofuels, enzymes, and fine chemical
2	Lignocellulosic waste	• Lignin and tannins • Sugarcane bagasse • Straw dust • Tea waste • Coconut waste • Wheat straw • Fruit waste	Animal feed, composites, perfumes, pieces of stuff, pulp and paper, and biochar
3	Municipal waste	• Sewage waste • Industrial waste	bio-alcohol and biofuels

4.3.1 Sugary Waste

All over the world, more than 30 million tons of sludge from the sugar industry are generated per annum, out of which India generates 12 million tons. The sludge is stored mostly in open fields by the sugar mills/industries, which degrade that area of land and water. The sugar industry generates wastes such as bagasse, bagasse fly ash, molasses, sugar beet pulp, sugarcane trash, press mud, and sugar beet mud. These sugar wastes in combination with various other organic materials get converted into perfect mixtures for the earthworm's growth. If these wastes are stored in the open area, they may cause environmental pollution that can result in several diseases to public health. This led to the development of technologies that are environmentally friendly and cost-effective for recycling and/or bioremediation of nutrient wastes, which may help in reducing the pollutants generated and conserving resources and the natural environment of the ecosystems.

Bagasse is the fibrous residue, which is left over from sugarcane after juice has been extracted. Bagasse is used as a mulch or soil conditioner, cattle feed, poultry litter, and organic manure. It is also used for the production of ethanol, fuel, and mushroom, and the preparation of biochar. Sugarcane bagasse fly ash is obtained after the burning of bagasse, which is used as fuel in the sugar industry. It is used as an absorbent for heavy metal removal, manure, and treatment of wastewater. The production of molasses is done from the separation of crystals of sucrose after evaporation and during the sugar crystal production from clarified juice of sugarcane or beet. It is a viscous liquid of dark chocolate color and contains almost 15–20% water, 50–55% of total sugars, and the rest non-sugars. It is used as animal feed, manure, silage preservative, insect bait, fertigation, and reclamation of soil. Sugar beet pulp is a fragment of fine slices of sugar beets (cassettes), exhausted by hot water and pressed. The exhausting of the cassette corresponds to the diffusion process achieved in sugar factories to extract the accumulated sugar from the sugar beet in its root. The cane trash is produced during sugar processing, and its quantity depends on the plant diversity, harvest time, age of the crop, humidity, and characteristics of soil and weather conditions. The total above-ground biomass at harvest is 15%, which is equal to about 10–15 tons/hectare of the dry mass of matter.

The press mud is an organic waste in the form of filter cake. This press mud is generated from sugarcane industries as a by-product. It is characterized as a soft material, spongy in texture, and amorphous powder, with dark brown or brownish color material. It is generated during the recovery of sugar in the form of filter cake generated by the carbonation or sulfation process. It is used as organic manure, carp culture, animal feed, organic amendment, and carrier for legume inoculants and substrate for bio-composting. Similarly, sugar beet mud is a sugar industrial sludge/solid waste in the form of filter cake formed after the recovery of sugar by using carbonation or the sulfation process. This is a two-step process in which it separates clear juice on top and filters cake/sludge at the bottom of the separator. Vermistabilization is one of the alternative technologies in which the earthworms are used for the utilization of sludges generated from the sugar industry and to convert them into a valuable/profitable product such as organic manure. This is the low-cost bioconversion technology that will detoxify the toxic chemicals present in sludge and will convert it into less or nontoxic and usable forms with the help of the vermiremediation prospective and potential of earthworms. The detoxification of sludges/mud in the sugar industry by using earthworms or other microbes will provide a sustainable and suitable alternative method to manage industrial solid waste and produce a profitable product along with solving the growing pollution problem.

4.3.2 Lignocellulosic

Lignocellulosic biomass (LCB) is a natural resource that is renewable and economical for use in production via bioprocess techniques all over the world. The major components of LCB are lignin, hemicellulose, and cellulose. The separation of these components of LCB is very difficult, and it is a major challenge to develop proficient, efficient, and cost-effective methods. The fractionated LCB is converted into monomeric substrates, which are utilized for biochemical production. LCB is also classified as waste material generated from agriculture, like corn stover, energy crops like switchgrass, woody residue, and human waste. LCB can be used as a raw material for biotechnology and chemical companies and industries. The applications of LCB are having various technical challenges for industries. Various studies are carried out on lignocellulosic material, fungal consortia, and ligninolytic enzymes such as lignin peroxidase, manganese peroxidase, and laccase to confirm their potential for the application of biotechnology [6, 7]. The biorefinery potentially utilizes almost all kinds of biomass sources, which include agricultural waste, agro-industrial waste, algae, sewage, and municipal resources. Biorefinery is normally based on sugar fermentation, extracted from various feedstocks based on polysaccharides that are from the natural source. These biofuels reduce the dependency on petrochemical products. It prevents the increase in petroleum prices and reduces the overconsumption of coal, gas, petroleum, and other potential minerals. These biofuels enhance and diversify biorenewable sources and solve the problems associated with global climate and greenhouse gas emissions. It also boosts regional and rural greener growth and safeguards the natural ecosystem. Based on the selection of raw materials (waste material) and processing techniques, different types of biorefineries were developed [8]. A green biorefinery system has great potential to use green biomass for the production of various products such as fibers, fuel, animal feed, and ethanol [9, 10]. A wide range of lignocellulosic polysaccharide substrates generated naturally or through agro-industries is available for the bioprocess industry to produce high-value products. The bioconversion of such substances satisfies the demands of industry [11–13] and also solves the pollution problem by converting them to zero waste.

4.3.3 Municipal Waste

Municipal waste is defined as waste collected, treated, and managed by municipalities. Municipal solid waste is material generated from homes, businesses, and industries. The waste from households, institutions, office buildings, small-scale businesses, domestic yards, garden waste, street sweepings, litter, market cleaning waste in containers, commerce, trading companies, and industries are the various types of municipal waste. The waste generated from construction and demolition activities is also municipal waste and is also known as trash or garbage. The waste includes food materials, papers and pulp, plastic materials, textile materials, leather parts, wood and straw dust, glass particles, metals and ores, sewage and sanitary waste in containers and septic tanks, and present in many other places.

In the distillery industry, the production of 1 L of alcohol generates about 15 L of wastewater. The generated raw spent wash is dark brown and acidic with an unpleasant odor. Its COD is 100,000 mg/L and BOD 45,000 mg/L, which is almost very high. This spent wash is used for the production of methane gas through the biomethanation process. In India, almost 1,100 million cubic feet of methane gas per annum can be produced using waste spent produced by the distillery industry. Spent wash can also be used for cattle feed supplements, fertigation, soil ameliorant, and vermicompost.

Biochar is a solid product that is rich in the percentage of carbon content. The adsorption properties of biochar are ideal for wastewater treatment, which can be utilized for several environmental applications. Biochar along with wastewater treatment is also applicable in power generation, microbial fuel cell production, fuel production, dye, and other pollutant removals. It is an economically and environmentally sustainable process. Many more such innovative bioengineering technologies are effective tools for the treatment of such type of waste material and convert the waste into a valuable product.

4.4 Production Process and Instruments

The waste is utilized as a substrate through various stages of the production process for product formation. In the process of product formation, various instruments are used for the production and recovery of products. Some more instruments are also used for giving proper shape and packaging to the product so that they are acceptable in the market for purchase.

4.4.1 Bioprocess Production Layout for Waste Utilization

The steps involved in bioprocess include the selection of organisms, inoculums/seed preparation (growth of microorganism), selection of substrate for microbial growth and product formation, production of a product by utilizing waste material, and recovery process of the product as shown in Figure 4.3. The biocatalysts (enzymes) or immobilized enzymes are also used for the bioconversion of organic waste into the product. Crushing, grinding, chemical/biological pretreatment, and drying of waste materials are the various techniques that are used to enhance the utilization of waste and finally increase productivity during the production process. The fermentation process is used for product formation. The various steps are involved in the recovery of the product, such as removal of insoluble, isolation, purification, and polishing of the product. These recovery processes involve various techniques like filtration, centrifugation, adsorption, extraction, protein precipitation, chromatography, crystallization, and drying. One or more techniques can be selected for the utilization of waste and product formation. These processes are scaled up to the pilot plant and then finally scaled up to the industrial level.

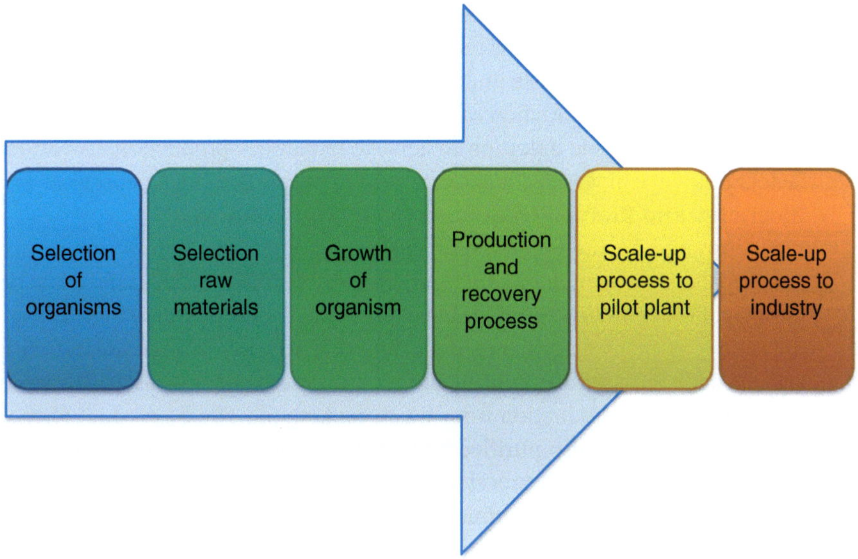

Figure 4.3 Bioprocess steps for the production of the product.

4.4.2 Bioprocess Instruments for Production and Recovery

The production process for the utilization of waste to produce desired products requires bioreactors and other instruments. Various types of bioreactors are involved in the process, such as anaerobic membrane bioreactors for biohydrogen production, Koji fermenters for solid-state fermentation for the production of different chemicals, photobioreactors for the production of algal biofuel, and many more. The various recovery equipment is used for various production processes such as distillation assemblies, dryers, mixers, crystallizers, evaporators, etc. The distillation assemblies and evaporators are used for the production of perfume, oils, and essential oils from agricultural waste and waste flowers, etc.

4.5 Bioprocess Application to Achieve Zero Waste

The few bioprocess applications for the utilization of various waste materials to achieve zero waste are discussed in this section.

- **Biodegradable Plastics Production from Agriculture and Food Waste**
  The agriculture and food waste material are utilized for the production of biodegradable plastics and bioplastic packaging materials from waste material to achieve zero waste. The waste utilized is sugar and starch waste, dairy and milk waste, and other wastewater.

- **Biogas Production from Waste**
  Biogas is also bioenergy that is produced from waste using an anaerobic biorefinery approach. Food waste and municipal waste are utilized for the production of biogas. The production of biogas involves pretreatment and variations in microbial anaerobic digestion to achieve maximum production, as various factors influence biogas production and the application of metagenomics.
- **Bioenergy and Biofuel Production from Waste Material**
  Bioenergy and biofuel production is done by utilizing agricultural waste, municipal wastewater, waste cooking oil, etc. The microalgae easily grow on municipal wastewater to produce algal biofuel. The production of biofuel from this source is many fold higher than the other sources. Waste cooking oil is also used for the production of biodiesel. The process of transesterification is used for biodiesel production from waste cooking oil. The by-product formed is glycerol, which is also purified and used in various industries such as cosmetics, pharma, etc. This process also leads to the zero-waste concept and is economical. Zero-waste biorefineries are a bioeconomic strategy around the world. Although various challenging factors impact the bioeconomy, research is still in progress.
- **Biohydrogen Production from Waste Biomass**
  Biomass is used as a feedstock for the production of biohydrogen. The factors affect biohydrogen production, so the parameters are optimized to achieve the highest yield. Various strategies can be applied to enhance microbial hydrogen production.
- **Bioethanol Production from Lignocellulosic Waste**
  Bioethanol is produced from lignocellulosic waste using microbes with the help of an anaerobic and aerobic production process. The LCB is pretreated by hydrolyzing with the help of acid, alkali, or enzyme and then converted into ethanol using microbes such as *Saccharomyces cerevisiae* by the fermentation process.
- **Biobutanol Production from Starchy and Lignocellulosic Waste as Residues**
  Biobutanol is produced from starchy and lignocellulosic waste as residues to achieve zero waste. The production process involves microorganisms that convert sugars present in waste material into butanol with high yields via fermentation technology.
- **Compost Production from Solid Waste**
  The compost is produced from solid waste using microbes and their consortia. Various bacteria, fungi, insects, and worms are microbes that play a very important role in the formation of composting. The microbial consortia are much more effective for the production of solid waste composting. Vermicomposting is also well known as one such technology in which earthworm synergizes with microbial degradation to reduce, reuse, and recycle waste materials in a short time duration. Composting is a process in which microorganisms grow on organic matter and break down it to produce gas like carbon dioxide, humus, water, heat, and a stable organic end product. The process of composting is

done in three phases under optimal conditions. The optimum conditions are the moderate-temperature phase, or mesophilic phase, which takes a couple of days, the high-temperature phase, or thermophilic phase, which takes a few days to several months, and the third phase, which takes several months for cooling and maturation. There are so many benefits of microbial compost, such as containing a full spectrum of essential macro- and micronutrients for the plant. These nutrients are absent in synthetic fertilizers but are the need for specific plants. Compost buffers, neutralizes, and enriches the soil to the optimize range of nutrient availability to plants. Compost helps in loosening tightly bound particles in silt and clay soil; as a result, the roots can spread, water gets easily drained, and air penetrates at a fast rate. All of this helps sandy soil retain water and nutrients and also alters soil structure. The advantage of compost is that it reduces the use of chemical pesticides and contains beneficial microorganisms. It also protects plants from diseases and pests.

- **Nanomaterials Biogeneration from Agriculture, Horticulture, and Food Wastes**
  Nanotechnology and nanobiotechnology are novel and versatile biotechnology approaches for the removal of promising contaminants to convert them into zero waste. The green synthesis of nanomaterials is done by fermentation of food, agricultural, and horticultural waste. The nanoparticle is characterized and applied in various fields such as biopolymer, biofilm, and bio-coating. The application of nanotechnology is also used for immobilized enzymes and bioconversion.
- **Edible Coatings and Films Produced from Marine Food and Agricultural Wastes**
  The agricultural and agro-industrial residues are used for the production of biologically active compounds, such as bio-derived polymers. The agro-industrial residues such as starch and starch derivatives, pectin, collagen, gelatin, chitin, chitosan, and fats are used for the production of films and coatings. Recently nanocomposites area is also used for improving biopolymeric film and coating properties. The main nanoparticles obtained from agro-industrial residues are cellulose nanofibers based on LCB, such as wheat straw, soy hulls, rice straw, sugarcane bagasse, cassava bagasse, sisal, pea, hemp, flax, and others, and also chitosan nanoparticles from marine food processing or bacterial origin. Still, the toxicity and environmental impact of nanocomposites need more study.

4.6 Bioconversion of Waste to Valuable Product a Bioeconomy Approach

These techniques of bioconversion enhance entrepreneurship that can promote the utilization of waste into a profitable product. This technique is a bioeconomic approach, as there are various approaches available to reduce, reuse, and process the waste generated to convert it into useful and valuable products at a very low cost.

These bioprocess techniques help in bioeconomic waste utilization for production and management.

4.7 Profitable Entrepreneurship to Achieving Zero Waste

The rivers are the main source of water, food, bathing, and agriculture for millions of people. In India, most of the holy rivers are used for workshops, baths, and dumping of temple flowers, which are polluting rivers. The temple flower waste material is dumped every year on the land and water bodies in tons. As India is a religious country, these temple flower wastes are of major concern because these waste materials are increasing pollution on land and water and are killing aquatic animals and other living beings. Fertilizers and pesticides present in the flowers are also polluting the land and water bodies. Also, the temple waste takes longer to decompose than expected. This problem leads to thinking about the solution, and if the solutions are profitable, then it would be bonus points.

- In India, the showering of flowers at temples/mosques due to religious rituals is almost 8 million tons annually. These sacred flowers create disaster in the fragile ecosphere of the water bodies and groundwater because they are mostly released in the River Ganga or nature. This pollution problem was observed by Ankit Agarwal and Karan Rastogi, the founders of this green company, with the trademarked Flowercycling®. They started the company HelpUsGreen® in 2015 in a 21,000 sq. ft. area, which is based in Uttar Pradesh, India. The company was a profitable company with full-time employees and around 73 women providing them all the benefits, such as health insurance, provident funds, and transport to and from work. The company recycles floral waste and converts it into useful products such as biodegradable packaging material, charcoal-free incense, and organic compost. The company recycled 11,060 tons of flowers and eliminated 11 tons of pesticides, and achieved impactful results. The company collects 8.4 tons of floral waste daily from temples in Uttar Pradesh and offers a profitable solution to the waste generated from the temple.

Pesticides, weedicides, and insecticides are used to grow flowers in gardens and farms. When these flowers are mixed with water bodies, they make them highly poisonous. The pollution caused by these flowers causes diseases among water animals and other populations and causes death [14].

The company adopted the process that consists of stages as follows:

- Flower wastes are brought to the company after it is collected from the temples.
- The segregation of flowers is done by hand to weed out plastics or papers from it.
- The organic bio-inoculums are sprayed on the flowers to offset the chemical residue; this bio-inoculum used is developed by the company.
- After this, the flowers are washed carefully and then stored and used in composting.

The flower petals and their parts are broken and dried in sun rays. The carotenoid level and other characteristics of flowers are checked depending on the type of species of the flowers. The company adopts three different processes to produce environment-friendly incense sticks: the organic compost known as mitti, composed of a mixture of 17 natural ingredients, including discarded coffee grounds by the local outposts of a coffee chain, and the innovative floral foam as a product for packaging purposes. Initially, the traditional thermocouple was used as packaging materials, which are leaving a negative impact on the environment. The functionality of Florafoam® is superior to Thermocol; it is also an eco-friendly and cheaper alternative to Thermocol. It is 100% biodegradable and fire-resistant. This biodegradable floral foam can be made of any shape and size. Their strength can be maintained depending on their usability. The company extended its plant in Varanasi, Mathura, and Tirupati (under construction). These new plants will recycle 12 tons of flower waste a day and will employ 128 women. The target of the company was to stop the ending of flowers in the river so that the pollution can be controlled [14].

- Maya Vivek and Minal Dalmia, two enterprising women, started a startup with the name Holy Waste, which manufactures products out of floral waste. It was started in December in Gundlapochampally, over 20 km from Hyderabad city. They used the flowers from temples and turned them into useful everyday products such as the sweet fragrance of soaps and incense sticks [15].

Similarly, various other wastes are also polluting the environment in various ways, such as vegetable waste, e-waste, etc. The vegetable waste is utilized with the help of bioprocess technology to convert it into profitable products such as biofuel, compost (vermicompost, bio manure, co-composting), fodder and feed, and various metabolites (enzymes, organic acids, etc.). These waste materials can also be utilized for the process of bioremediation [16]. The waste can also be utilized for the production of biodegradable stationery, disposable cups, glasses, plates, and various other products. These innovative ideas can be of great help to zero-waste concepts. The utilization of waste into a valuable product is economic, profitable, socially, and environmentally beneficial.

4.8 Conclusion

It is difficult and will take time to understand the waste problem that is directly or indirectly damaging the environment and can cause havoc in upcoming years. Now it is high time to explore ideas and technologies that can utilize these wastes to reduce pollution and develop profitable products. The new products can be more developed by utilizing waste to convert it into zero waste. It is time to develop and implement novel technologies to utilize waste and come up with the problem of waste instead of waiting and trying to develop awareness among the people against the negative impact of waste and pollution. It is time to search for an applicable and accessible alternative to get the waste solutions and achieve zero waste. Life without waste and living with 100% zero waste is next to impossible, but together, we can approach getting zero waste along with valuable products. More and more research is required in the field of bioprocess technologies to achieve the target and develop the profitable business of biological products using waste as a substrate of zero cost to approach zero-waste targets.

References

1. Palmer, P. (2004). *Getting to Zero Waste*. Sebastopol, CA: Purple Sky Press.
2. LaBrecque, S. (2015). *Talkpoint: What Does Zero Waste Mean to You*. Manchester, UK: The Guardian.
3. Greyson, J. (2007). An economic instrument for zero waste, economic growth and sustainability. *J Clean Prod*. 15 (13–14): 1382–1390.
4. Zaman, A.U. (2015). A comprehensive review of the development of zero waste management: lessons learned and guidelines. *J Clean Prod*. 91: 12–25.
5. Gupta, M., Savla, N., Pandit, C., et al. (2022). Use of biomass-derived biochar in wastewater treatment and power production: a promising solution for a sustainable environment. *Sci Total Environ*. 825: 153892.
6. Iqbal, H.M., Kyazze, G., and Keshavarz, T. (2013). Advances in the valorization of lignocellulosic materials by biotechnology: an overview. *BioResources*. 8 (2): 3157–3176.
7. Arevalo-Gallegos, A., Ahmad, Z., Asgher, M., et al. (2017). Lignocellulose: a sustainable material to produce value-added products with a zero waste approach—a review. *Int J Biol Macromol*. 99: 308–318.
8. Parajuli, R., Dalgaard, T., Jørgensen, U., et al. (2015). Biorefining in the prevailing energy and materials crisis: a review of sustainable pathways for biorefinery value chains and sustainability assessment methodologies. *Renew Sustain Energy Rev*. 43: 244–263.
9. Scholey, D.V., Burton, E.J., and Williams, P.E. (2016). The bio refinery; producing feed and fuel from grain. *Food Chem*. 197: 937–942.
10. Karlsson, H., Ahlgren, S., Strid, I., and Hansson, P.A. (2015). Faba beans for biorefinery feedstock or feed? Greenhouse gas and energy balances of different applications. *Agric Syst*. 141: 138–148.
11. Sakamoto, T., Hasunuma, T., Hori, Y., et al. (2012). Direct ethanol production from hemicellulosic materials of rice straw by use of an engineered yeast strain codisplaying three types of hemicellulolytic enzymes on the surface of xylose-utilizing Saccharomyces cerevisiae cells. *J Biotech*. 158 (4): 203–210.
12. Shahsavarani, H., Hasegawa, D., Yokota, D., et al. (2013). Enhanced bio-ethanol production from cellulosic materials by semi-simultaneous saccharification and fermentation using high temperature resistant Saccharomyces cerevisiae TJ14. *J Biosci Bioeng*. 115 (1): 20–23.
13. Usmani, Z., Sharma, M., Awasthi, A.K., et al. (2021). Bioprocessing of waste biomass for sustainable product development and minimizing environmental impact. *Bioresour Technol*. 322: 124548.

14 Recycling floral waste in India [Internet] (2023). https://www.ideassonline.org/public/pdf/IndiaRecyclingFlowers-ENG.pdf (accessed 15 January 2024).

15 Turning floral waste into soap, incense: The story of a women-run Telangana startup [Internet] (2023). https://www.thenewsminute.com/telangana/turning-floral-waste-soap-incense-story-women-run-telangana-startup-106119 (accessed 15 January 2024).

16 Singh, A., Kuila, A., Adak, S., et al. (2011). Use of fermentation technology on vegetable residues for value added product development: a concept of zero waste utilization. *Int J Food Ferment Technol.* 1 (2): 173–184.

5

Biological Processes and Techniques for Valorization of Household Wastes

5.1 Introduction

The global generation of household waste has reached unprecedented levels due to rapid urbanization, industrialization, and rising living standards. According to the World Bank's "What a Waste 2.0" report, the world produces over 2.01 billion tonnes of municipal solid waste annually, with at least 33% of it not managed in an environmentally safe manner. This figure is projected to increase to 3.4 billion tonnes by 2050 if current trends continue. A substantial portion of this waste originates from domestic activities, including food preparation, packaging, cleaning, and lifestyle consumption. Household waste typically comprises organic matter (such as food scraps and yard trimmings), paper and cardboard, textiles, plastics, glass, and small quantities of hazardous waste such as batteries and cleaning chemicals. The composition of household waste varies across regions, largely influenced by income levels, dietary habits, and consumption patterns. Low-income countries tend to generate higher proportions of organic waste, while high-income nations produce more packaging, plastics, and electronic waste. For instance, in low-income areas, food and garden waste can constitute up to 60–70% of household waste, whereas in high-income regions, nonbiodegradable components such as plastic and composite packaging are more prevalent. This diversity in composition presents both challenges and opportunities for waste management strategies, especially when aiming for sustainable and efficient recovery of resources. The environmental and economic implications of untreated household waste are severe. Open dumping and unregulated landfilling remain common practices in many parts of the world, particularly in developing nations. These methods contribute significantly to environmental degradation, including soil and groundwater contamination, air pollution from methane and leachate emissions, and the proliferation of disease vectors. Incineration, while reducing the volume of waste, can release hazardous substances such as dioxins and furans if not properly controlled. The failure to adequately treat domestic waste also leads to the loss of potentially valuable materials that could be recovered or reused,

Biological Processes and Techniques in Zero Waste Management, First Edition. Garima Awasthi and Mohit Nigam. Published 2026 by Wiley-VCH GmbH.

representing a significant missed economic opportunity. Economically, the burden of waste management often falls on municipal governments, many of which lack the financial and technical capacity to implement modern waste treatment facilities [1]. The costs associated with landfill maintenance, waste collection logistics, and environmental remediation are substantial. Moreover, the informal sector, which plays a critical role in recycling and material recovery in many developing countries, often operates under unsafe and unsanitary conditions, lacking formal recognition and support. The economic inefficiency of the linear "take-make-dispose" model not only strains public resources but also squanders materials that could be part of a more circular and regenerative economic system. In response to these challenges, the concept of waste valorization has gained global traction. Waste valorization refers to the process of transforming waste materials into useful products, including energy, compost, bio-based chemicals, and new materials. It is a core principle of the circular economy, a model that seeks to decouple economic growth from resource consumption by emphasizing reuse, recycling, and regeneration. In this context, household waste is no longer seen as a disposal problem but as a resource stream that, if managed properly, can contribute to sustainable development goals (SDGs), energy security, and environmental preservation.

The biological valorization of household waste, in particular, offers an innovative and sustainable pathway. By leveraging microbial and enzymatic systems, organic fractions of domestic waste can be converted into biofertilizers, biogas, biofuels, and other value-added products. This not only mitigates environmental harm but also creates economic opportunities through the development of green technologies and local circular economies. Thus, rethinking household waste management through the lens of biological valorization is essential for a sustainable future [2].

5.2 The Need for Biological Valorization

Conventional methods of household waste management, particularly landfilling and incineration, are becoming increasingly unsustainable in the face of growing urban populations, limited land availability, and escalating environmental concerns. Landfilling, while cost-effective in the short term, leads to long-term issues such as leachate contamination, methane emissions, a potent greenhouse gas, and loss of potentially valuable resources. On the other hand, incineration reduces waste volume but contributes significantly to air pollution and the emission of harmful compounds like dioxins and furans. Both methods emphasize waste elimination rather than recovery, failing to align with modern principles of circular economy and resource efficiency. In contrast, biological valorization presents an environmentally sound alternative that integrates well with sustainability frameworks. This approach utilizes biological processes, such as composting, anaerobic digestion (AD), and microbial fermentation, to transform organic waste fractions into valuable products like biofuels, bioplastics, organic fertilizers, and enzymes. These methods not only minimize the environmental footprint of waste disposal but also enhance the recovery of energy and nutrients. Biological valorization processes are typically less energy-intensive and more adaptable to decentralized waste management systems, which are critical for rural and peri-urban areas. Moreover, from a life cycle perspective, biological methods offer multiple sustainability benefits. They facilitate the diversion of organic waste from landfills, reduce greenhouse gas emissions, and support soil health through compost application. In addition, biological treatments can be designed to integrate with existing waste management systems and support cascading use of biomass, whereby waste is progressively transformed into products of increasing value. This enhances resource circularity and aligns with the principles of zero-waste frameworks. The global policy landscape increasingly favors such sustainable waste management strategies. The United Nations SDGs, especially SDG 11 (Sustainable Cities and Communities), SDG 12 (Responsible Consumption and Production), and SDG 13 (Climate Action), emphasize the need for inclusive, sustainable, and resilient urban development, in which waste valorization plays a central role. National and regional regulations are also evolving to promote waste segregation, composting, and decentralized treatment solutions. Incentive-based mechanisms and policy support for bioeconomy initiatives are further catalyzing the adoption of biological processes for waste valorization. Biological valorization also contributes significantly to climate resilience and resource recovery. By reducing dependence on fossil fuels and chemical fertilizers through the production of bio-based alternatives, it mitigates climate impacts and enhances local resource security [3]. Furthermore, it allows for adaptive, scalable solutions that can be customized to local contexts, making it particularly relevant in developing countries where infrastructure and resource constraints limit the effectiveness of conventional waste treatment systems. The transition toward biological valorization is not only a scientific and technical shift but also a socio-environmental necessity. As we face increasing pressures to decarbonize economies and close material loops, biologically driven waste valorization stands out as a promising strategy for sustainable development, ecological restoration, and circular economy advancement.

5.3 Categories of Household Wastes Amenable to Biological Valorization

The growing emphasis on sustainable waste management has led to increased interest in the biological valorization of household waste. Biological valorization refers to the transformation of biodegradable waste into valuable products through processes such as composting, AD, fermentation, and enzymatic conversion. Among the diverse components of municipal solid waste, several categories generated in households are particularly suitable for biological treatment due to their high organic content and degradability. This section highlights the major categories of household waste that can be effectively valorized through biological means [4].

5.3.1 Organic Kitchen Waste

Organic kitchen waste constitutes a major fraction of household refuse and includes vegetable peels, fruit scraps, leftover food, tea leaves, eggshells, and used cooking oils. These wastes are rich in biodegradable organic matter and serve as excellent feedstocks for composting and AD. Their high-moisture content and balanced carbon-to-nitrogen (C/N) ratio support microbial activity, enabling rapid degradation and conversion into compost or biogas. For instance, food scraps can be digested anaerobically to produce methane-rich biogas, while the resulting digestate can be used as organic fertilizer. Used cooking oil, though not biodegradable in the conventional sense, can be transformed through microbial or enzymatic processes into biodiesel or biopolymers. Proper segregation of kitchen waste at the source enhances the efficiency and quality of biological valorization processes.

5.3.2 Paper and Lignocellulosic Residues

Paper waste, including used napkins, cardboard, newspapers, and packaging paper, is predominantly composed of cellulose, hemicellulose, and lignin, collectively known as lignocellulosic biomass. Though these materials are slower to degrade than kitchen waste, they are valuable feedstocks for bioconversion due to their abundance and carbon-rich composition. Biological pretreatment methods, such as microbial or enzymatic hydrolysis, can break down complex lignocellulosic structures into fermentable sugars. These sugars can then be fermented by microorganisms to produce bioethanol, organic acids, and other biochemicals. Additionally, shredded paper can be co-composted with food waste to improve aeration and balance the C/N ratio, leading to higher quality compost [5].

5.3.3 Garden and Yard Waste

Yard waste, including grass clippings, leaves, pruned branches, flowers, and weeds, is another significant category amenable to biological valorization. These materials are often seasonal but present in large volumes in suburban and urban households. They contain high levels of fibrous cellulose and lignin, making them suitable for composting and bioenergy production. The fibrous nature of garden waste helps in improving the structure of compost piles by facilitating airflow and reducing compaction. Composting garden waste results in nutrient-rich humus that can be used to improve soil health. Woody residues, although slower to decompose, can be shredded and co-composted or used in mushroom cultivation after appropriate treatment. In some advanced systems, yard waste is also utilized in producing biogas or biochar, depending on the technology employed [6].

5.3.4 Low-grade Biodegradable Plastics and Packaging Materials

Recent advances in material science have led to the introduction of biodegradable plastics such as polylactic acid (PLA), polyhydroxyalkanoates (PHAs), and starch-based polymers in household packaging. These materials, when properly identified and segregated, can be subjected to controlled composting or microbial degradation. Although their degradation is slower than that of food or garden waste, they can be effectively broken down under industrial composting conditions or with the help of specialized microbial consortia. The valorization of such plastics not only reduces the burden of synthetic plastic waste but also contributes to circular bioeconomy strategies. However, proper labeling, consumer awareness, and supportive policy mechanisms are crucial for the successful integration of these materials into biological waste valorization streams [7].

5.3.5 Textile Residues of Natural Origin

Household textile waste, especially from worn-out clothes, towels, or upholstery made of natural fibers like cotton, linen, silk, and wool, represents another underutilized category of biodegradable waste. These fibers are primarily composed of cellulose or proteins and can be degraded biologically, albeit at a slower rate than kitchen or garden waste. Enzymatic hydrolysis and microbial degradation techniques can convert cotton and wool into simpler compounds, which may be further processed into bioethanol, organic acids, or compost. Though often mixed with synthetic fibers in modern textiles, natural fiber residues can be separated and valorized using appropriate technologies. Incorporating textile waste into composting processes can also help in moisture absorption and improving compost texture when combined with high-moisture organic materials [8].

5.4 Biological Processes and Techniques

The biological valorization of household waste presents a sustainable and eco-friendly alternative to conventional disposal methods. These techniques leverage microbial and enzymatic systems to convert biodegradable waste into valuable end products such as biofuels, bioplastics, organic acids, and compost. This section details the fundamental biological processes and technologies involved in the valorization of organic household waste.

5.4.1 Composting

Composting is a natural aerobic process where organic waste is biologically decomposed by microorganisms, resulting in a nutrient-rich soil conditioner. It is one of the most accessible and widely used techniques for treating kitchen waste, garden residues, and other biodegradable household materials.

Microbial succession during composting is a key aspect of the process. Initially, mesophilic bacteria thrive at moderate temperatures and begin breaking down simple organic compounds. As the temperature rises due to microbial activity, thermophilic bacteria and fungi dominate, rapidly degrading more complex materials like cellulose and lignin. Once most of the organic matter is decomposed, the temperature decreases, allowing mesophilic organisms to recolonize and stabilize the compost.

Parameters affecting compost quality include the C/N ratio, which should ideally be between 25:1 and 30:1 for optimal microbial activity. Aeration is crucial to maintaining aerobic conditions, preventing foul odors, and supporting microbial respiration. Moisture content should be kept between 50 and 60% to ensure adequate microbial metabolism without promoting anaerobic conditions [9].

In-vessel composting is a controlled method where composting occurs in enclosed reactors, offering enhanced temperature regulation and pathogen control. Vermicomposting, on the other hand, utilizes earthworms (e.g., *Eisenia fetida*) to further decompose organic material, resulting in highly nutritious vermicast with superior agronomic value.

5.4.2 Anaerobic Digestion

AD is a biological process conducted in the absence of oxygen, where microorganisms break down organic matter to produce biogas (mainly methane and carbon dioxide) and digestate, a nutrient-rich slurry [10].

The process involves four biochemical stages:

1. **Hydrolysis:** Large organic molecules such as proteins, fats, and carbohydrates are broken down into simpler compounds like amino acids, fatty acids, and sugars.
2. **Acidogenesis:** These simpler compounds are converted into volatile fatty acids, alcohols, and gases like hydrogen and carbon dioxide.
3. **Acetogenesis:** Further breakdown of intermediate products into acetic acid, hydrogen, and carbon dioxide.
4. **Methanogenesis:** Methanogenic archaea convert acetic acid and hydrogen into methane and carbon dioxide.

Reactor types vary depending on the nature of the waste and operational scale. UASB (Upflow Anaerobic Sludge Blanket) reactors are suitable for high-strength wastewater with suspended solids. CSTR (Continuous Stirred Tank Reactor) systems are ideal for homogeneous slurries and maintain uniform mixing. Plug flow reactors are suited for high-solids feedstocks and provide longer retention times for slow-degrading materials.

Biogas upgrading involves removing impurities like carbon dioxide, hydrogen sulfide, and water vapor to increase the methane content, making it suitable for energy generation or injection into the gas grid. The digestate can be dewatered and applied as a biofertilizer, closing the nutrient loop.

5.4.3 Enzymatic Hydrolysis and Bioconversion

Enzymatic hydrolysis is a process that breaks down complex polysaccharides present in lignocellulosic biomass into simple fermentable sugars using specific enzyme systems. These sugars are then used in various bioconversion processes to produce industrially relevant products [11].

Enzyme systems used for lignocellulose degradation include:

- **Cellulases:** Break down cellulose into glucose units.
- **Xylanases:** Act on hemicellulose to release xylose and other sugars.
- **Ligninases:** Aid in the partial breakdown of lignin to improve cellulose access.

Due to the recalcitrant nature of lignocellulose, pretreatment strategies are essential. These may include:

- Alkaline pretreatment (e.g., NaOH or ammonia) to remove lignin and swell fibers.
- Acid pretreatment (e.g., H_2SO_4) to hydrolyze hemicellulose.
- Steam explosion or enzymatic preconditioning to increase surface area and porosity.

Applications of enzymatic hydrolysis products include:

- Biosurfactants, such as rhamnolipids, which are biodegradable and eco-friendly alternatives to synthetic surfactants.
- Bioplastics, such as PHAs, which are produced by bacteria from sugar monomers.
- Organic acids like lactic acid and succinic acid, which serve as precursors for biodegradable polymers and solvents.

5.4.4 Microbial Fermentation for Bio-product Recovery

Fermentation is a microbial process where organic substrates are converted into valuable bio-products under anaerobic or aerobic conditions. Household waste-derived substrates like sugar-rich hydrolysates, waste oils, and starches are suitable for microbial fermentation.

Fermentation can be carried out using [12]:

- Wild-type microbes, such as *Saccharomyces cerevisiae* (ethanol), *Clostridium* spp. (butanol), or *Aspergillus niger* (citric acid).
- Genetically engineered microbes, which are optimized for higher yield, faster growth, or tolerance to toxic substrates.

Products recovered through fermentation include:

- Ethanol and butanol, used as biofuels or solvents.
- Organic acids such as lactic, citric, or acetic acid, used in food, pharma, and chemical industries.
- Pigments, like carotenoids and prodigiosin, produced by microbes for use in cosmetics and nutraceuticals.
- Biosurfactants and microbial lipids, produced from waste cooking oils or fatty substrates. These have applications in detergents, bioremediation, and bio-lubricants.

The use of fed-batch or continuous fermentation systems allows for enhanced product yield and process stability, especially in industrial setups.

5.4.5 Co-culture and Synthetic Biology Approaches

Recent advancements in biotechnology have introduced co-culture systems and synthetic biology approaches to optimize the biological valorization of complex waste streams. Microbial consortia involve multiple microbial species working synergistically to degrade heterogeneous waste materials. For instance, cellulolytic bacteria

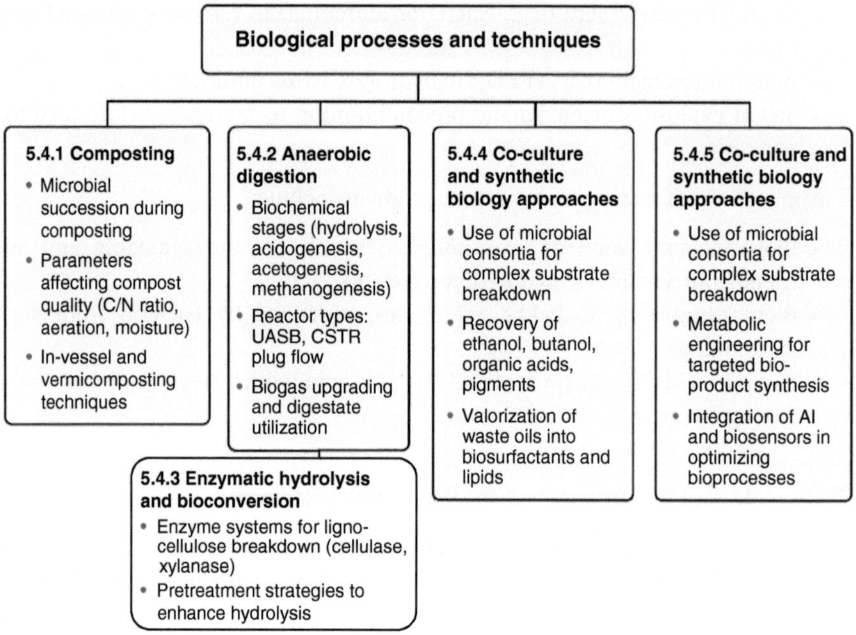

Figure 5.1 Overview of key biological processes and techniques for waste conversion and resource recovery.

may degrade fibers into sugars, while fermentative microbes convert those sugars into ethanol or acids [13].

Metabolic engineering involves the modification of microbial genomes to redirect metabolic flux toward the production of desired compounds. Techniques such as CRISPR-Cas9, gene knock-in/knockout, and pathway reconstruction are used to design strains capable of:

- Producing non-native compounds.
- Enhancing tolerance to inhibitors.
- Increasing substrate uptake rates.

Integration of AI and biosensors enhances process monitoring and control. Biosensors can detect pH, dissolved oxygen, or product concentration in real time. Machine learning models analyze large datasets from fermenters or digesters to optimize feeding rates, aeration, and enzyme dosing. These tools are pivotal in creating intelligent biorefineries that are efficient, adaptable, and economically viable for waste valorization at scale (Figure 5.1).

5.5 Case Studies and Global Best Practices

Biological valorization of household waste has emerged as a sustainable and scalable solution for waste management across the globe. From decentralized composting initiatives in India to sophisticated biogas production systems in European cities, real-world examples demonstrate how tailored biological approaches can turn household waste into economic and environmental assets. This section explores notable case studies and best practices from around the world, emphasizing their methodologies, success factors, and potential for replication in other contexts [14].

5.5.1 Community-based Household Composting in India

India, with its vast population and diverse socio-economic landscape, faces enormous challenges in managing household waste. In many urban and semi-urban areas, community-based composting initiatives have proven effective in handling organic waste at the source.

One notable example is the Swachh cooperative in Pune, Maharashtra, where waste pickers are integrated into the formal solid waste management system. Organic waste from households is collected, sorted, and composted at decentralized locations within residential societies. This system uses simple composting bins and pit composting methods, supported by minimal mechanical infrastructure, making it cost-effective and replicable [15].

Key features of this model include:

- **Source Segregation Education:** Residents are trained to separate wet and dry waste, ensuring cleaner input for composting.
- **Microbial Composting Additives:** Natural inoculants like cow dung slurry or microbial culture are used to accelerate decomposition.
- **Income Generation:** Compost is sold to local farmers or used in community gardens, and waste pickers earn a livelihood from compost sales and user fees [16].

Benefits observed:

- Reduction in municipal waste transportation and landfill burden.
- Employment and empowerment of informal waste workers.
- Community ownership and awareness of sustainable practices.

Similar programs have been implemented in cities like Bengaluru (with Hasiru Dala) and Chennai, where housing societies take collective responsibility for managing organic waste through aerobic composting and vermicomposting [17].

5.5.2 Biogas from Kitchen Waste in Urban Europe

Several European cities have adopted innovative and technology-driven methods for converting kitchen and food waste into biogas, primarily through AD. These systems are often integrated with municipal waste collection and energy supply networks.

A successful example is the Aarhus Municipality in Denmark, which operates a centralized biogas plant that processes food waste collected from households and restaurants. Waste is first pretreated to remove packaging and contaminants before undergoing AD in CSTRs [18].

Key technological elements:

- Automated waste sorting and de-packaging units.
- High-efficiency AD reactors with biogas yield optimization via real-time monitoring.
- Biogas upgrading units, which refine the raw biogas into biomethane suitable for injection into the natural gas grid or use in city buses.

Another noteworthy example is Stockholm, Sweden, where the city's food waste collection system is closely linked to its public transport. The biogas produced is used to fuel buses, creating a circular link between households and mobility services.

Impact and outcomes:

- Significant reduction in landfill dependence and GHG emissions.
- Decentralized energy production contributing to renewable energy targets.
- Enhanced public participation through incentives and awareness campaigns.

These European models demonstrate how kitchen waste can be transformed from a disposal problem into a local energy resource, supported by strong policy frameworks, investment in infrastructure, and public engagement.

5.5.3 Decentralized Valorization Units for Smart Cities

In the context of smart city development, decentralized biological valorization units are being increasingly recognized for their flexibility, scalability, and low environmental footprint. These systems are especially suitable for rapidly urbanizing regions with limited land availability and strained municipal waste management systems [19].

India's Smart City Mission has seen the rise of several pilot projects that deploy compact composting and biogas units within residential colonies, commercial complexes, and institutional campuses. Technologies include:

- In-vessel composting units with odor control mechanisms.
- Microbial inoculum-based rapid composters.
- Modular biogas digesters capable of handling 100–500 kg of organic waste daily.

A successful example is the NEERI-designed "BIOMETHANE" plant installed in Bhopal Smart City. The unit digests segregated food and vegetable waste to produce biogas, which is used for street lighting and small-scale cooking applications, while the slurry is converted into organic fertilizer.

Advantages of decentralized units:

- Minimized waste transportation costs and emissions.
- Customization based on local waste characteristics and volumes.
- Engagement of local entrepreneurs and communities in operations.

In other global examples, cities like San Francisco (USA) and Freiburg (Germany) have adopted zero-waste strategies that include neighborhood-scale composting and fermentation systems as integral components of urban sustainability.

5.5.4 Industrial Partnerships and Upscaling Efforts

As the value of biological waste valorization becomes more apparent, public–private partnerships (PPPs) and industrial collaborations are playing a key role in scaling up these solutions.

One example is the collaboration between Veolia Environment and local municipalities in France, where large-scale AD plants treat both household and agro-industrial organic waste. These facilities produce biogas, heat, and nutrient-rich compost, supported by advanced logistics, monitoring, and regulatory compliance systems. In Japan, the food retail chain AEON partners with local technology providers to recycle food waste into energy and fertilizer through integrated composting and AD plants located behind large retail stores. This closed-loop approach supports corporate sustainability goals and reduces overall carbon footprints [20].

- Strong government incentives and regulatory support for waste segregation and renewable energy.
- Investment in R&D for improving digestion and composting efficiency.
- Market development for compost, biogas, and derivative products.
- Training and capacity-building for technical staff and waste handlers.

Emerging industrial models also include biorefineries that co-process various organic wastes into multiple bio-products (e.g., enzymes, organic acids, biochar), creating diversified revenue streams. These integrated facilities are gaining traction in Europe and North America, supported by circular economy policies and green investment mechanisms.

5.6 Technological Challenges and Solutions

- **Feedstock Variability and Pretreatment Issues**
  One of the major technological barriers in biological waste valorization is the heterogeneity of household waste feedstock. Organic waste from homes often varies in composition, moisture content, particle size, and presence of inert or nonbiodegradable materials. This inconsistency impacts the efficiency and predictability of biological processes such as composting, AD, and fermentation. To address this, proper segregation at the source is essential. However, even with effective segregation, the complexity of feedstock requires advanced pretreatment technologies. Mechanical shredding, homogenization, and screening are commonly used to ensure uniformity. In addition, chemical or thermal pretreatment methods, such as steam explosion, dilute acid treatment, and alkali treatment, are used to break down complex polymers like lignin and cellulose, particularly in lignocellulosic biomass. Developing region-specific protocols and investing in modular pretreatment systems can enhance process readiness and efficiency [21].
- **Contamination and Pathogen Control**
  Household organic waste often contains contaminants such as plastics, metals, glass, and hazardous substances. Moreover, the presence of pathogens poses

health risks and can hinder the microbial activity essential for biological valorization processes. These contaminants must be identified and removed before processing to ensure both safety and process efficiency. Contamination control can be achieved through improved source segregation practices, coupled with mechanical sorting systems equipped with sensors and artificial intelligence (AI)-based vision technologies. Pathogen control, especially in composting and AD, requires maintaining appropriate temperature and moisture conditions. Thermophilic composting (55–65 °C) and pasteurization steps in digestion are effective in reducing pathogens. Ensuring hygienic operation protocols and periodic microbial analysis can further minimize risks [22].

- **Reactor Scalability and Cost Limitations**
  Scaling up biological waste treatment systems from pilot to industrial levels presents numerous engineering and economic challenges. Reactor design must account for variable waste volumes, retention times, and microbial kinetics, while ensuring operational stability and minimal energy input. Compact and modular reactor systems have been proposed as a solution to scalability, particularly for decentralized settings. However, these often come with higher initial investment costs. CSTRs, plug flow digesters, and in-vessel composters require robust infrastructure, skilled operators, and reliable monitoring equipment. Cost-effective reactor designs must strike a balance between capacity, durability, and automation. Innovations in low-cost sensors, remote monitoring, and mobile-integrated control systems can help reduce operational expenses and make small-scale units more accessible [23].
- **Integration with Existing Municipal Waste Systems**
  For biological valorization to be successful on a large scale, it must be seamlessly integrated into existing municipal waste management infrastructure. However, aligning decentralized biological units with centralized collection and disposal systems can be complex. Municipal bodies often face logistical and administrative challenges in restructuring waste collection routes, enabling decentralized composting or digestion units, and monitoring compliance. Policy support, including incentives for segregated waste generation and financial models for PPPs, can help facilitate this integration. Digital tools like GIS-based waste tracking, data-driven optimization of collection logistics, and community engagement platforms are increasingly being adopted to support this integration. Ultimately, a systems approach that includes technological, regulatory, and social components is required for successful integration of biological processes within broader urban waste management frameworks.

5.7 Future Trends and Research Directions

- **CRISPR and Synthetic Biology in Waste-to-value**
  Advancements in genetic engineering, particularly CRISPR-Cas systems and synthetic biology, are revolutionizing the field of biological waste valorization.

These technologies allow precise manipulation of microbial genomes, enabling the development of engineered strains with enhanced metabolic capabilities. Such modifications can improve the efficiency of bio-product synthesis from household waste and expand the range of feedstocks that can be utilized.

CRISPR-based gene editing has enabled the development of microbial strains that can break down complex polymers more effectively, tolerate inhibitors found in waste streams, and produce higher yields of desirable products such as bioethanol, lactic acid, or bioplastics. Synthetic biology goes a step further by constructing entirely novel metabolic pathways or microbial consortia, designed to function synergistically under specific conditions.

For example, synthetic microbial consortia can be programmed to degrade mixed organic waste into valuable precursors like acetyl-CoA, which can be further converted into fuels, surfactants, and fine chemicals. As these technologies mature, their integration into decentralized biorefineries holds promise for scalable, high-efficiency waste valorization solutions [24].

- **AI-based Bioprocess Optimization**

  AI is playing a growing role in optimizing biological waste conversion processes. Traditional bioprocesses often suffer from variability and are sensitive to environmental fluctuations. AI and machine learning tools can analyze large datasets from sensors and laboratory experiments to predict outcomes, optimize operating parameters, and reduce human error.

  Applications of AI in bioprocessing include predictive modeling of microbial growth, enzyme activity, and fermentation kinetics. Real-time monitoring systems equipped with biosensors and linked to AI algorithms can dynamically adjust pH, temperature, aeration, and nutrient dosing to maximize yields. This results in more stable, efficient, and cost-effective processes.

  Moreover, digital twins, virtual replicas of physical systems, are being used to simulate bioprocessing environments for testing and optimization before implementation. These advancements support the development of intelligent, adaptive systems suitable for managing the complexity of mixed household waste streams [25].

- **Decentralized Biorefineries for Household Waste**

  The concept of decentralized biorefineries involves small- to medium-scale processing units that convert household organic waste into bio-based products at or near the point of generation. These biorefineries aim to reduce the logistical burden of waste transport and promote circular economy principles within communities.

  Modern decentralized systems combine several technologies, including AD, enzymatic hydrolysis, microbial fermentation, and composting. They can produce biogas, compost, organic acids, and value-added chemicals locally, supporting energy self-sufficiency and resource recovery at the neighborhood level.

  Advances in modular design, automation, and remote monitoring have made decentralized biorefineries more viable. They are particularly relevant for urban areas where centralized waste processing is logistically difficult and rural areas where energy access is limited. These systems can be tailored to community needs and scaled over time, providing a flexible and resilient waste management solution. The successful implementation of future waste valorization technologies depends heavily on enabling policy frameworks and economic incentives. Governments play a critical role in fostering innovation through funding, subsidies, and regulations that promote the adoption of biological waste treatment technologies.

Policies supporting source segregation, extended producer responsibility (EPR), and zero-waste targets create favorable conditions for the growth of waste-to-value industries. Incentives such as feed-in tariffs for biogas, carbon credits, and startup grants for decentralized systems can accelerate deployment and commercialization.

In addition, integrating waste valorization goals with national climate action plans and SDGs ensures long-term institutional commitment. PPPs, community participation, and awareness campaigns are also essential in building the social infrastructure needed for success.

Future research should focus on life cycle assessments, techno-economic analyses, and inclusive policy instruments that address economic, social, and environmental dimensions of household waste valorization.

5.8 Conclusion

Biological valorization of household waste represents a frontier of sustainable innovation where environmental stewardship meets biotechnology. By employing microbial, enzymatic, and consortia-based strategies, even low-value domestic residues can be transformed into high-impact products. The shift toward biologically driven waste management is not only necessary but inevitable, provided technological, economic, and policy frameworks evolve in synergy. With growing urban populations and increasing pressure on resources, biologically valorizing household waste will be essential for achieving circular economy goals, ensuring energy security, and safeguarding the environment.

References

1 Awasthi, M.K., Pandey, A.K., Bundela, P.S., and Khan, J. (2014). Co-composting of organic waste materials: a review. *Int J Environ Res Dev*. 4 (4): 371–376.

2 Bernal, M.P., Alburquerque, J.A., and Moral, R. (2009). Composting of animal manures and chemical criteria for compost maturity assessment. *Bioresour Technol*. 100 (22): 5444–5453.

3 Bolzonella, D., Battistoni, P., Susini, C., and Cecchi, F. (2006). Anaerobic co-digestion of waste activated sludge and OFMSW: the experiences of Viareggio and Treviso plants (Italy). *Water Sci Technol*. 53 (8): 203–211.

4 Campanaro, S., Treu, L., Kougias, P.G., and Angelidaki, I. (2018). Taxonomy of anaerobic digestion microbiome reveals biases associated with the applied high throughput sequencing strategies. *Sci Rep.* 8 (1): 1926.

5 Capson-Tojo, G., Rouez, M., Crest, M., et al. (2020). Strategies to limit inhibition during dry anaerobic digestion: a review. *Bioresour Technol.* 313: 123607.

6 Chandel, A.K., Garlapati, V.K., Singh, A.K., et al. (2018). The path forward for lignocellulose biorefineries: bottlenecks, solutions, and perspectives on commercialization. *Bioresour Technol.* 264: 370–381.

7 Chen, H. and Liu, Y. (2007). Enzymatic hydrolysis of lignocellulosic biomass. *Bioresour.* 2 (4): 929–955.

8 Demichelis, F., Fiore, S., Pleissner, D., and Venus, J. (2021). Food waste valorization via fermentation: a review of the technological approaches. *J Environ Manag.* 288: 112416.

9 Diaz, L.F., Savage, G.M., and Eggerth, L.L. (2005). Composting and compost quality assurance. *Waste Manag.* 25 (6): 501–504.

10 Donoso-Bravo, A., Mailier, J., Martín, C., et al. (2011). Model selection, identification and validation in anaerobic digestion: a review. *Water Res.* 45 (17): 5347–5364.

11 Escobar, S., García, C., and Guerrero, Y. (2020). AI in anaerobic digestion process: review and perspectives. *Renew Sustain Energy Rev.* 132: 110045.

12 Kavitha, S. and Kumar, M.D. (2020). Decentralized biorefineries for urban waste valorization: a circular economy approach. *Clean Waste Syst.* 1: 100002.

13 Khatami, K., Aghbashlo, M., Tabatabaei, M., and Wongwatanapaiboon, J. (2021). The role of synthetic biology in bioenergy: opportunities and challenges. *Renew Sustain Energy Rev.* 135: 110415.

14 Li, Y., Park, S.Y., and Zhu, J. (2011). Solid-state anaerobic digestion for methane production from organic waste. *Renew Sustain Energy Rev.* 15 (1): 821–826.

15 Makarova, K.S., Wolf, Y.I., and Koonin, E.V. (2018). The basic biology of CRISPR–Cas systems and their role in biotechnology. *Annu Rev Microbiol.* 72: 447–465.

16 Mohan, S.V., Modestra, J.A., Amulya, K., et al. (2016). A circular bioeconomy with biobased products from CO_2 sequestration. *Trends Biotechnol.* 34 (6): 506–519.

17 Montusiewicz, A. and Lebiocka, M. (2011). Influence of co-substrate on anaerobic digestion of sewage sludge. *Environ Technol.* 32 (8): 897–907.

18 Nasr, M., Tawfik, A., Ookawara, S., et al. (2012). Modeling and optimization of hybrid anaerobic digestion systems for kitchen waste treatment. *Bioresour Technol.* 114: 90–98.

19 Pandey, A., Soccol, C.R., Nigam, P., and Soccol, V.T. (2000). Biotechnological potential of agro-industrial residues. *Bioresour Technol.* 74 (1): 81–87.

20 Patnaik, S. and Das, D. (2021). Biorefinery approach for household organic waste management: technological and policy dimensions. *Waste Manag Res.* 39 (6): 727–736.

21 Rouches, E., Achinas, S., and Euverink, G.J.W. (2021). Fermentation technologies for valorizing household biowaste. *Biotechnol Adv.* 49: 107751.

22 Shah, A.A. and Jan, M.R. (2018). Composting of organic waste using microbial inoculants: recent advances and future prospects. *J Environ Manag*. 217: 108–119.

23 Singh, A. and Bishnoi, N.R. (2013). Enzymatic hydrolysis of pretreated rice straw by cellulases from *Aspergillus heteromorphus. Bioresour Technol*. 135: 180–184.

24 Srinivasan, R., Kumar, S., and Ghoshal, A.K. (2020). Decentralized waste-to-energy systems: policy drivers and barriers in India. *Renew Energy*. 156: 892–903.

25 Zhi, Z., Jiang, M., and Sun, Y. (2022). Integration of AI and biosensors for real-time control in microbial fermentation. *Biosyst Eng*. 215: 14–24.

6

Scientific Approach for Valorization of Industrial Waste

6.1 Introduction

In this modern era of industries, there are many sources of production and extraction of goods and chattels, which constantly have an extended, unsatisfying, and ever-growing culture of consumption [1]. There is a very wide range of consumer goods and products, like clothes, white goods, and many electronic products, which are treated as luxury items that are used and consumed as everyday goods in the present scenario [2]. The production process is now transformed vigorously into a complex system that mainly consists of composite and hazardous materials. As a result, today we are producing a lot of waste from many mixed sources that damage the environment and are also very expensive to manage. There are a variety of wastes produced nowadays, which has left the decision makers to choose these inefficient waste management methods that damage the environment continuously, such as a landfill. Therefore, the scarcity of landfill sites in urban areas has now forced the waste management authorities to look up an alternative method for the efficient management of waste produced [3]. Waste management has now a visionary system introduced as zero-waste approach, which offers an alternative option and solution for every waste-related problem in the present scenario [4]. Zero waste is now turned into an inspirational and ultimate goal to deal with the waste problems. There are many cities like Vancouver, San Francisco, and Adelaide, which have adopted the policy of zero waste as a part of their management strategies [4]. It is a concept that is highly embraced by the policymakers that help in stimulating effective and sustainable production as well as consumption of produced goods and also optimizes the resource recycling and recovery while simultaneously restricting the mass incineration and landfilling of waste.

6.1.1 Role of Industrial Manufacturing in Attaining Zero-waste Concept

Manufacturing is an essential part of sustaining growth in the economic sector. Gross Domestic Product (GDP) plays an important role in manufacturing, ranging from 15 to 25% of national economies. A quarter of worldwide employment is accounted for via manufacturing itself. In the present scenario of industrialization, further expansion has been observed in the goods manufacturing and services to fulfill the global demand [5]. Nowadays, a wide range of products is there in the market, which comprises domestic need items, remarkably advanced electrical equipment, vehicles, tool manufacturing and design, edibles, etc. [4]. Generally, one has to sacrifice a huge number of resources for the production of these varieties of products that are available now in the market easily, as these processes are extremely complex and costly. Inclusively, the point is that these used resources are not able to recover or be reused efficiently, hence, contributing directly to manufacturing waste. It can be observed clearly that the source of this manufacturing waste is not at all specific in nature, and it can be of any category of process or material, yet it is mutual for all. As a serious repercussion of this issue, the whole world is now facing the grave problem of generated waste from comprehensive manufacturing activity [6] (Figure 6.1).

Curran and Williams (2012) have clearly defined the main aim of the zero-waste system approach entirely, which is to "eradicate" rather than just "manage" waste [7]. Zero-emissions processes represent a further step away from the outdated industrial manufacturing practices where wastes are taken up from one system and then used as input material for another. This directly supports an industrial-level alteration that can diminish the impact of waste on natural resources.

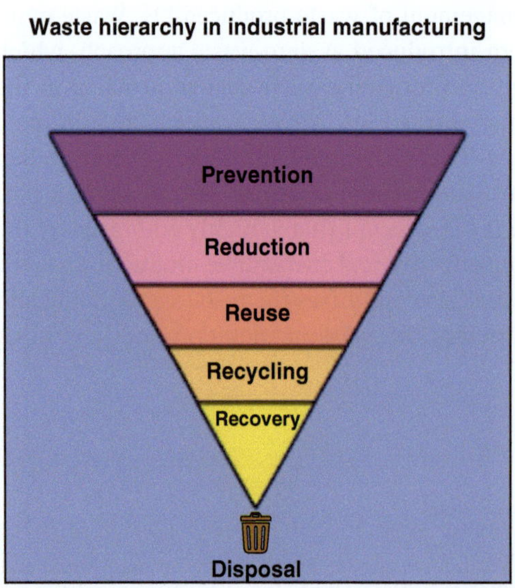

Figure 6.1 A pyramid diagram showing the waste hierarchy.

Zero waste can be classified into the following sets of systems:

- Zero waste under administrative and manufacturing units.
- Zero waste of natural and other significant resources.
- Zero discharges.
- Zero waste throughout the life cycle of a product.
- Zero utilization of toxins in manufacturing processes.

The waste cannot be dumped without treatment because it is unhealthy, unhygienic, and potentially devastating to our environment and mother earth. It requires the allocation of proper place and space and also entails consequences and costs related to waste disposal. The sites for landfill are also very difficult to find for dumping. The process cost is also very high for identifying the location and responsible landfill facilities. The waste is generated at a very high rate, and soon it will not be easy to manage the dumping sites and locations for landfill. It was reported in 2009 that waste produced by Malaysia, USA, Europe, Singapore, Vietnam, Thailand, Brunei, Cambodia, and Myanmar is 0.81, 2.2, 1.5, 1.1, 0.61, 64, 52, 0.45, and 0.66 kg/cap/day, respectively [8].

Production and manufacturing play an important role in the generation of waste and can be solved using various techniques and methodologies, in which the things can be utilized multiple times. The manufacturers must be aware of producing multiuse products [9]. Also, we need to develop various techniques and technologies for utilizing waste and converting it into useful products to achieve zero waste.

6.2 Scientific Pathways Employed in Different Industrial Sectors for Zero-waste Discharge

It is observed that a huge amount of waste is generated around the globe due to heavy industrialization and urbanization. Also, the variety of waste is dependent on the type of industry and community living standards. Zero-waste discharge is a concept that can be applied to all types of waste generated from these industries, and the waste can be utilized as a resource for creating or manufacturing some other product/process or can be reduced by following new-age recycling and waste-reducing techniques. Some of these major industries and their zero-waste adoption practices are discussed further in this chapter.

6.2.1 Leather Industry

Various chemicals are being used in leather processing to convert the rawhides/skin into the final leather product. There are different types of waste generated by the leather industries. Rawhide processing from the animals produces many solid wastes, such as flesh, hair, chrome trimmings, and various other solid wastes during the process. The effluents which are released from the leather industries have a very higher concentration of toxic chemicals and heavy metals which further degrades the environment critically. Therefore, to control the contaminants being released into the environment and subsequently reduce waste generation, leather industries are now adapting various modern technologies on the production side and also in the management of effluents. Further, depending upon the type of waste or effluents involves various procedures are now adopted in the effort toward zero-waste generation. There are several modern technologies viz., bioremediation, bioprocessing technology, enzyme technology, microbial technology, and various advanced treatment technologies for effluents [10].

6.2.2 Cleaner Preservation of Skin/Hide Using Silica Gel

The cleaner conservation technique using the silica gel method proposed the conventional salt curing method. It suggested that the preservation method is cost-effective, easy to adapt, and an environment-friendly approach. A better preservation system can be conventional, using inert compound silica gel. This novel method uses the silica gel, which comprises sodium meta-silicate and hydrochloric acid at a pH of 5.5, and is further followed by spray drying. The resultant powder is used for preservation purposes. The recommended preservation method used an optimal amount of silica gel powder at 15% with or without 0.1% PCMC (Para Chloro Meta Cresol). Moving further, the less-salt method of consuming silica gel at the level of 5% with salt 5% has also been developed now. The preservation efficacy using the new method was edited by various parameters such as moisture content using Nuclear Magnetic Resonance (NMR), micro-imaging technique, total extractable nitrogen using nitrogen analyzer, and bacterial count by standard agar plate technique. The fiber structures were analyzed by using a method scanning electron microscope (SEM). Therefore, the results now indicate that the new cleaner preservation technique was much more effective than salt curing, and it did not cause any kind of problems during the soaking and manufacturing of leather. This preservation method was established to be a feasible alternative to reduce the pollution load in treated skins and combat the problems like Cl and TDS [11, 12].

6.2.3 Eco-friendly Method for Dehairing of Skin/Hide

The dehairing method is one of the main operations conducted in the pre-tanning process, where hair is removed by using suitable chemicals. The traditional method of liming is based on the application of lime, viz., 10e20%, and sodium sulfide, 2e5%, which produces a lot of pollution problems in the environment. This method generates approximately 50% of COD and 40% of BOD, which further contributes to 60–70% of the total pollution load in leather processing [13]. The global concerns about the environmental impact of the leather industry have now led tanners to reduce the components of toxicity in the effluents [14]. Nevertheless, sodium sulfide is a very toxic chemical that bears many unfavorable consequences to the environment and the efficacy of effluent treatment plants. As a remedy, there are alternate viable methods that are proposed by using protease to replace the lime and sodium sulfide for eco-friendly processing of leather [15, 16]. Researcher Senthilvelan et al. (2012) stated that enzymatic unhairing can be done by using protease enzymes to replace sulfide [17]. Protease enzyme was obtained from *Bacillus* sp. The optimum activity of the protease enzyme was found at pH 11.0, at a temperature of 45 °C, and for a duration of 16 hours. As a result, it was found that the complete dehairing was achieved by using alkaline protease at 2% for 16 hours. This method provided the sustainable technology where activities of BOD, COD, TSS, and TDS were further reduced at the level of 62.8, 79.0, 88.2, and 82.5%, respectively, in evaluation with the control sample [18, 19].

6.2.4 Vegetable Tanning

The leather tanning is of chrome-tanning type; there are improved cleaner tanning options, which have been practiced and performed in the case of vegetable-tanned leather. Vegetable tanning is one of the traditional tanning systems, which comprises natural plant materials such as leaves, barks, nuts, and fruits to extract tannins (polyphenols) for tanning of the skin or in hides. Researcher Kanth et al. (2009) talk about cleaner production options using natural products during leather processing to reduce the COD and TDS in the product [20].

6.2.5 Textile Industry

The textile industry is a huge pollution source because it produces raw constituents (natural and synthetic fibers), preparation and final procedures, and also textile waste, particularly postconsumer waste. In reference to the circular economy, textile-related waste must be regarded as a carbon and energy source. A novel mindset is obligatory because of the surge in the quantity of postconsumer waste, subsequently after the increase in the volume of textile consumption. Cycles of fast fashion, as well as economical textile products with a brief lifetime, have led to an upsurge in the amount of postconsumer textile waste. For reducing the carbon and water footprints of these textile products, intense demands have been made for a reduction in pollution by at least somewhat consuming the textile waste for recovering its constituents and the energy consumed for their manufacture. These demands involve the valorization of postconsumer textile waste via recycling and partially recovering the materials and energy resources used in product manufacturing. For reaching the milestone of zero-waste production, we need an excellent postconsumer textile waste management strategy with compulsory implementation [21, 22].

Valorization of textile waste can be achieved through the following methods:

- Recycling textile products as second-hand clothes or completely remanufacturing them into fresh garments and pieces of decoration. This can be one of the best ways to get the most out of any textile waste. Clothes having longer life is useful in reducing raw material consumption and energy use, which in turn reduces pollution, which results from the production of these textile products.

Therefore, a decrease in the carbon and water footprints of textile products can be seen noticeably, leading to sustainable growth.

- Textile waste upcycling to get fibers, nonwoven material, and composites, as well as using it to acquire various polymers or organic compounds, are other valued resolutions. The newly obtained materials can be successful in replacing wood or metals, thus declining their consumption.
- Another useful application of textile waste can be using it as an energy source in place of fossil fuels, hence, declining the fossil combustible depletion.
- Textile waste can also be applied in cultivated lands for property reinforcement as well as recovery.

6.2.6 Food Processing Industry

The food produced, which is meant for human consumption but is wasted every year, is nearly 1.3 billion tons worldwide, which sums up to around one-third of the entire production. Food loss and waste is one of the root causes of grave social, economic, and environmental problems discouraging the sustainability of our planet. Thus, actions pointed at food loss and waste prevention as well as reduction should be taken with urgency. In this context, the move toward a circular economy model implemented in this area is now very urgent to encourage the conversion to a better sustainable agro-food structure, stirring toward a zero-waste industry.

6.2.7 Source Reduction

The biggest environmental and materials efficiency will come from not wasting food at all. Approaches and characterizations of source reduction hinge on the phases of the food supply chain taken into consideration and are deeply tied to the characteristics of food waste. If we consider "food" only as something edible for humans, then several by-products of food production, like peelings, food trimmings, meat and bones, or evoked food products, will not be considered for waste reduction at source, yet they can still be used for material recovery (e.g., fertilizer made out of bones) as well as energy recovery (e.g., by-products of vegetables utilized as a food source in an anaerobic digester) [23].

6.2.8 Feeding Hungry Population

For situations in which surplus food qualifies the set safety and health criteria, then instead of dumping it in landfills, it can be provided to people in need. Two vital bits of regulations have been decreed by the United States to boost contributions of eatable foods to people who require food. First is the Bill Emerson Good Samaritan Food Donation Act, which aims to defend contributors and beneficiaries from legal accountability, and the second is the Protecting Americans from Tax Hikes Act 2015, donating food enables suitable food sector businesses to claim any possible deductions in the aspect of the cost of the food that is donated [24, 25]. Yet, numerous administrators are either uninformed of these bits of the legislature and are too averse to the potential risk involved to follow food donations, or have a problem with investing time to follow these possibilities, as they do not find it valuable [21].

6.2.9 Feed Animals

One other possibility is to make food scrapings available to the livestock directly to feed on, but in the maximum number of cases, supplementary treatment is essential to confirm the safety of animals and to diminish the danger of foodborne diseases to humans. As is already known, the most likely candidates for livestock feed are the remaining organic materials from farmhouses or food processing units. Also, postconsumer food waste used as feedstocks is very consistent and poses a low risk in regard to contamination possibilities [26].

6.2.10 Industrial Uses

Food waste for industrial uses is an important category, and it generally includes the production of value-adding products produced from food waste, with the main attraction towards oil recovery to be exploited as biodiesel. The abstraction of value-adding items mainly happens by the use of food processing by-product chains, instead of postconsumer food waste. This is because food processing units have a better and very homogeneous nature of food waste or by-product stream. It is more favorable to consequent engineered extractions of precise composites. One of the examples is the phenolic compounds, which can be utilized as an antioxidant in supplementary foods, essential oils, pectin, proteins, and several different important compounds [27].

6.2.11 Biomedical and Healthcare Industry

The biomedical waste comprises the healthcare units, like the waste from laboratories of medical institutions, and waste streams from biomedical research centers. Improper handling of this waste elevates many grave hazards of ailment spread caused by coming in contact with the infectious materials for the biomedical medical waste handling personnel, the healthcare workers, the already sick and weak patients, and the common population [28]. Biomedical medical waste management has been increased further now during the pandemic of COVID-19 chiefly because of the surge in the biomedical medical waste volume, and this waste poses a significant threat to the environment and public health. The life cycle assessment and circular economy employed in the biomedical segment can further take on the waste management from sources like medical institutions, the pharmaceutical sector, and dentistry and can propose many techniques for "Green Circulation" of the generated waste.

The treatment of medical waste can be classified as

- Medical waste treatment on a large scale
  - **Incineration:** It is the destruction of waste material by burning, so it is combined with the heat recovery system. This process is mostly applied for the

treatment of medical waste to produce around 30 MJ/kg of heat and power, which is up to 80% recovery efficiency.

- **Pyrolysis:** It's a devolatilization process by the thermal decomposition of waste materials at high temperatures in an inert atmosphere. This process is applied to the plastic fraction of the medical waste to produce solid, gas, liquid, and fuel with higher heating values of 23 MJ/kg, 46 MJ/Nm^3, and 38 MJ/kg, respectively [29].
- **Rotary Kiln Treatment:** It is used in solid or liquid waste incineration.
- **Microwave/Steam Sterilization:** Such sterilization is an emerging technology; it is a very effective combination of microwave energy and low heat for the sterilization of medical waste. The microwave technology can be combined with autoclaving with sterilization steam at 93–177 °C for the killing of pathogens [30].
- **Plasma Gasification/Melting:** It is the most promising technique using plasma bioreactor for waste treatment at elevated temperatures and emissions of less hazardous materials. Plasma is the fourth state of matter and consists of charged particles, free electrons, ions, and radicals. The plasma gasification technology is used for H_2 and syngas production from medical waste [27].

• Medical waste treatment on the laboratory scale

- **Acid Hydrolysis:** It is a technique in which acids or combined acids are used to convert biomass into monomers. This process can be used for the fermentation of medical cotton waste into bioethanol.
- **Enzymatic Hydrolysis and Enzymatic Oxidation:** It is a process in which cellulases are used to convert cellulose into fermentable reducing sugars. These sugars are further converted into ethanol using yeasts or bacteria [18, 19].
- **Anaerobic Digestion:** This technique is used to produce biogas from industrial recycled medical cotton waste for which the thermophilic anaerobic biodigestion process is used to improve the yield of biogas up to 92% [19].
- **Autoclaving:** Similarly to large scale, various medical waste and pathogens are treated by using a combination of autoclaving and microwaving.
- **Hydrothermal Carbonization/Treatment:** This technology was applied to PVC, which contained medical waste mixed with lignocellulosic biomass to produce solid fuel on both the laboratory scale and pilot plant scale [31].
- **Sulfonation Process:** The fibrous cellulose sulphate absorbent was produced from medical cotton waste. The sulfonation process was used to synthesize this adsorbent by treating medical cotton waste and using Calcium Sulfoaluminate (CSA) in a dimethylformamide (DMF) medium [29].
- **Batch Reactor Thermal Cracking:** These reactors and processes are used for the production of biofuels.
- **Torrefaction:** It is the process used for the depolymerization of biomass.

The treatment of large-scale medical waste involves the segregation of waste fractions such as plastics and lignocellulose material. Various large-scale medical waste treatment techniques that are alternatives to the incineration process can be applied for the specific fractions of medical waste, and those processes are very environmentally friendly as compared to the incineration process. Still, large-scale laboratory medical waste treatment techniques are in the experimental phase for the treatment of medical waste fractions such as cotton, paper, cardboard, and textiles, which are cellulosic/lignocellulosic fractions.

These wastes can be used as a feedstock for the production of the following products:

- **Materials:** Adsorbents, fermentable sugars, etc.
- **Gas Fuels:** Biogas, syngas, hydrogen, etc.
- **Liquid Fuels:** bio-oil, bioethanol, etc.
- **Solid Fuels:** Pellets, hydrochar particles, etc. [2].

6.2.12 Paper and Plastic Industry

The recycling and reuse of paper- and plastic-origin waste is theoretically a capable approach to accomplishing a zero-waste target. The waste streams generated by paper and plastic industries are very crucial to the recovery or recycling of complete waste material to achieve the zero-waste goal. The United States and Europe have reported peak retrieval of material generated from the paper and plastic waste flow, which is around 85 and 73%, respectively. This indicates that still 15 and 27% of paper and plastic waste are left to degrade, which is not yet consumed or recycled. This indicates that complete recycling of all paper and plastic waste is impossible.

Various techniques are applicable to substantially increase the rates of recovery of paper and plastic waste. Recycled papers are continuously used, whereas the quality of such recovered paper is very low, and possibly that is the major low acceptance for its extensive use. Modern sorting technologies are used for the quality improvement of waste material. Similarly, there are various difficulties in sorting out plastics from other similar materials, which affect and bound the amounts of plastic waste that are in need of recycling. Currently, many approaches are made for recycling municipal waste, such as designing a vigorous sorting technique that can further reach the necessities of zero-waste goals. During the previous decades, the use of recycled paper has increased up to 46%; still, in the immediate past 5 years, this increase in the rate of recycled paper use slowed down about 1.2%, giving an average rate of 67.2% use of recycled paper for communities studied here. This can be considered a strong sign of a probable upper limit to paper recycling. The main causes for this can be the existence of huge contamination, poor quality of pulp, and increased processing costs. We can see a similar pattern in the case of plastic recycling [22].

Further, the approaches of biotechnology and bioengineering can be applied to complete the degradation of paper and plastic waste. These bioprocess technologies utilize such waste material generated from the paper industry in the form of biomass with the help of microbes and convert them into bioproducts such as the enzyme cellulase. Similarly, more products can be formed to utilize such waste and convert it into zero waste. More research is required to optimize the production process for substrate utilization and desired product formation.

6.3 Vermicomposting as Zero-waste Strategy

Solid wastes generated billions of tonnes annually that are dumped without any treatment in the environment, the cost of economic loss, and social and environmental issues. The appropriate management of such a huge volume of waste generated is a giant task for the entire world. To manage the waste, researchers are searching and working on various novel, groundbreaking, and environment-friendly methodologies. Biotechnology approaches have various suitable techniques for the treatment of waste. They can reutilize the numerous elements present in waste into valuable and profitable final products, which are cost-effective as well. The vermicomposting technique is biologically very effective. It is a natural process of bio-oxidative decomposition resulting from the biochemical action of microorganisms and takes place under mesophilic conditions. This process is used for waste management in which the waste is degraded with the help of earthworms assisted by microbes. The different species of earthworms are used for the vermicomposting of various categories of waste. The joint activity of earthworms and microorganisms helps in converting waste into a finely crushed, homogenized, odor free, rich in nutrients, and humus manure is known as vermicompost. Earthworms digest this solid waste as substrate in their intestine system and change their physical and chemical characteristics by aiding organic component breakdown. The microbes that are available in the intestine system of earthworms support the biodegradation of solid organic waste. Vermicompost is considered to be an effective plant growth enhancer since it encompasses nutrients useful for plants, enzymes, growth-promoting hormones, and humic substances, and is rich in the microbial population. In several studies, vermicompost has been proven best as an organic fertilizer for the improved development and harvest of crops. Vermicomposting also aids in achieving the goal of a circular and profitable economy by altering waste into valuable goods, which are obligatory for the complete sustainable growth of a nation [23].

6.4 Conclusion

In the current scenario, our society is getting concerned and waking up not to litter and enhance waste. Now people are worried about the environment and trying to protect it to reduce the damage caused to our mother world. Unknowingly we all leave an environmental footprint behind, so we should at least try to keep it as small as possible. In past decades life was difficult but natural and eco-friendly; nowadays technology has offered easy life and facilities for our comfort, but along with that, it has also given us huge waste that is damaging our planet day by day. Science and technologies have rapidly changed the market of fashion trends, electronic devices, excess food, plastic, motor equipment, etc., in a short period. These industries have created havoc with unconsumed material as waste along with the consumable market product. All such waste materials are producing hazardous greenhouse gases and contribute to global warming. Everyone is already aware of the waste generated and its hazardous impact on the environment. It is an urgent requirement to give a response to nature and society that how we can address such problems urgently. The scientific approaches are the only tool that can help to reduce, recover, recycle, and reconvert the components of waste material to turn into zero waste. The huge amount of waste generated from food (raw and cooked), agricultural, industrial, and medical industries can be utilized with the help of scientific approaches, specifically bioprocess technologies, to convert them into a valuable product. The zero-waste approach is holistic in nature and aims to solve the problem of generated waste in the current scenario with the help of various scientific approaches. This chapter concludes that zero-waste approaches are still in the developmental phase; more and more research is required in this area to achieve our target of zero waste. Although governments and professionals have proposed and implemented various ideas, processes, plans, policies, strategies, and technologies to achieve zero-waste goals in various cities worldwide, we still need to develop more scientific technologies for achieving a zero-waste strategy that can be implemented practically.

References

1 Antoniadou, M., Varzakas, T., and Tzoutzas, I. (2021). Circular economy in conjunction with treatment methodologies in the biomedical and dental waste sectors. *Circ Econ Sust.* 1 (2): 563–592.

2 Aravindhan, R., Saravanabhavan, S., Thanikaivelan, P., et al. (2007). A chemo-enzymatic pathway leads towards zero discharge tanning. *J Clean Prod.* 15: 1217–1227.

3 Connett, P. (2007). *Zero Waste: A Key Move Towards a Sustainable Society.* New York, USA.

4 Costello, C. (2019). The concept of zero waste. In: *Saving Food* (ed. C.M. Galanakis), 369–391. Academic Press.

5 Crocker, R. (2013). From access to excess: consumerism, 'compulsory' consumption and behavior change. In: *Motivating Change: Sustainable Design and Behavior in the Built Environment*, 1e (ed. R. Crocker and S. Lehmann), 43–64. Routledge.

6 Abdel-Hamied, M., Abdelhafez, A.A., Ahmed, R.F., Abd-Alrahman, S.H. et al. (2024). Evaluation of some fungicides for inhibiting proteolytic fungi isolated from leather binding of a historical manuscript dated back to the Mamluk period. *Herit Sci.* 12 (1).

7 Curran, T. and Williams, I.D. (2012). A zero-waste vision for industrial networks in Europe. *J Hazard Mater.* 207: 3–7.

8 Das, A.K., Islam, M.N., Billah, M.M., and Sarker, A. (2021). COVID-19 pandemic and healthcare solid waste management strategy–a mini-review. *Sci Total Environ.* 778: 146220.

9 Dimos, K., Paschos, T., Louloudi, A., et al. (2019). Effect of various pretreatment methods on bioethanol production from cotton stalks. *Fermentation.* 5 (1): 5.

10 Dowlath, M.J.H., Karuppannan, S.K., Rajan, P., et al. (2021). Application of advanced technologies in managing wastes produced by leather industries—an approach toward zero waste technology. In: *Concepts of Advanced Zero Waste Tools* (C.M. Hussain), 143–179. Elsevier.

11 Galanakis, C.M. (2012). Recovery of high added-value components from food wastes: conventional, emerging technologies, and commercialized applications. *Trends Food Sci Technol.* 26 (2): 68–87.

12 Hassan, A.H., Neely, K.E., and Workman, J.L. (2001). Histone acetyltransferase complexes stabilize swi/snf binding to promoter nucleosomes. *Cell.* 104 (6): 817–827.

13 Ismail, Z.Z. and Talib, A.R. (2016). Recycled medical cotton industry waste as a source of biogas recovery. *J Clean Prod.* 112: 4413–4418.

14 Kanth, S.V., Venba, R., Madhan, B., et al. (2009). Cleaner tanning practices for tannery pollution abatement: role of enzymes in eco-friendly vegetable tanning. *J Clean Prod.* 17 (5): 507–515.

15 Kumar, K., Gupta, S.C., Baidoo, S.K., et al. (2005). Antibiotic uptake by plants from soil fertilized with animal manure. *J Environ Qual.* 34 (6): 2082–2085.

16 Lopez, R. (2017). The environment as a factor of production: the effects of economic growth and trade liberalization. In: *International Trade and the Environment*, 239–260. Routledge.

17 Muzaffar, H.S., Mohit, N., Kumar, S.L., and Garima, A. (2020). Kinetic study of dyes degradation by aspergillus niger in submerged fermentation. *Res J Chem Environ.* 24 (4): 16–24.

18 Nigam, M., Husain, S.M., and Awasthi, G. (2018). Microbial production and purification of cellulase enzyme on waste as a substrate. *Toxicol Int.* 25 (1): 1–6.

19 Nigam, M., Yadav, R., and Awasthi, G. (2021). In-silico construction of hybrid ORF protein to enhance algal oil content for biofuel. In: *Advances in Biomedical Engineering and Technology* (ed. A.A. Rizvanov, B.K. Singh, and P. Ganasala), 67–89. Singapore: Springer.

20 Oppong, D., Bryant, S.D., Rangarajan, R., et al. (2006). Application of molecular techniques to identify bacteria isolated from the leather industry. *J Am Leather Chem Assoc.* 101 (4): 140–145.

21 Parfitt, J., Barthel, M., and Macnaughton, S. (2010). Food waste within food supply chains: quantification and potential for change to 2050. *Phil Trans R Soc B.* 365 (1554): 3065–3081.

22 Public Health Alliance of Southern California, California Conference of the Directors of Environmental Health, The Center for Climate Change and Health (2018). *Safe Surplus Food Donation Best Management Practices: Guidance for Environmental Health Departments.* Retrieved from https://bit.ly/2yXzHAe.

23 Rivin, J., Miller, Z., and Matel, O. (2012). *Using Food Waste as Livestock Feed.* United States: University of Wisconsin System Board of Regents and the University of Wisconsin-Extension, Cooperative Extension.

24 Senthilvelan, T., Kanagaraj, J., and Mandal, A.B. (2012). Application of enzymes for dehairing of skins: cleaner leather processing. *Clean Technol Environ Policy*. 14 (5): 889–897.

25 Sharma, D.K., Bapat, S., Brandes, W.F., et al. (2019). Technical feasibility of zero waste for paper and plastic wastes. *Waste Biomass Valor*. 10 (5): 1355–1363.

26 Sharma, K. and Garg, V.K. (2019). Vermicomposting of waste: a zero-waste approach for waste management. In: *Sustainable Resource Recovery and Zero Waste Approaches* (ed. M.J. Taherzadeh, K. Bolton, J. Wong, and A. Pandey), 133–164. Elsevier.

27 Sun, Y. and Cheng, J. (2002). Hydrolysis of lignocellulosic materials for ethanol production: a review. *Bioresour Technol*. 83 (1): 1–11.

28 Thakur, N., Nigam, M., Tewary, R., et al. (2022). Drivers for the behavioral receptiveness and non-receptiveness of farmers towards organic cultivation system. *J King Saud Univ Sci*. 34: 102107.

29 Thanikaivelan, P., Rao, J.R., Nair, B.U., and Ramasami, T. (2003). Approach towards zero discharge tanning: role of concentration on the development of eco-friendly liming–reliming processes. *J Clean Prod*. 11 (1): 79–90.

30 Wen, L., Lin, C.H., and Lee, S.C. (2009). Review of recycling performance indicators: a study on collection rate in Taiwan. *Waste Manag*. 29 (8): 2248–2256.

31 Zaman, A.U. (2014). Identification of key assessment indicators of the zero waste management systems. *Ecol Indic*. 36: 682–693.

7

Biological Techniques for Treatment of Biomedical Wastes and Wastewater

7.1 Introduction

Medical treatment is essential for our lives, health, and general well-being. The waste produced by medical procedures, however, may be dangerous, poisonous, and even fatal due to the high risk of disease transmission. Sharps (hypodermic needles, knives, scalpels, etc.) and other hazardous and toxic components of waste from healthcare facilities that contain infectious, biomedical, and radioactive material pose a serious risk if they are not properly handled, disposed of, or if they are permitted to mix with other municipal waste [1]. The efforts made for overall municipal waste management are jeopardized by its tendency to promote the development of different diseases and vectors as well as its capacity to contaminate other nonhazardous or nontoxic municipal garbage [2]. Whether they unintentionally or intentionally search through various types of hazardous material while attempting to rescue goods that they may sell for reuse, garbage workers and rag pickers are often the ones that suffer the most. Yet, this kind of illegitimate and unethical reuse may be lethal or exceedingly harmful. Serious hazards to public health are posed by illnesses like cholera, the plague, TB, hepatitis (particularly HBV), AIDS (HIV), diphtheria, etc., in epidemic or even endemic forms. Regrettably, it is challenging to estimate the size of the issue or even the scope and variety of the danger involved in the absence of trustworthy and comprehensive data.

Yet with careful planning and control, the danger may be greatly reduced. According to studies, nonhazardous and nontoxic waste makes up around three-fourths of the total trash produced in healthcare facilities [3].

Any solid or liquid waste, including its container and any intermediate products, produced during the diagnosis, treatment, or vaccination of humans or animals is referred to as biomedical waste (BMW). Hospitals, nursing homes, clinics, labs, doctor's, dentist's, and veterinarian's offices, home health care facilities, and funeral homes are common sources of biological waste. To prevent injury to the general public, particularly to sanitation and healthcare professionals who are often exposed to biological waste as an occupational hazard, it must be carefully handled [4]. Thus far, relatively little has been learned about how medical waste might spread infections. The American Dental Association and the Centers for Disease Control issued alerts stating that proper protocol must be followed when disposing of medical waste. A health care management program must include proper waste processing, treatment, and disposal. The right procedure safeguards patients, healthcare workers, and the community [5].

According to some estimates, hazardous trash is 15% and infectious waste is 5%. Hence, the issue may be correspondingly lessened with a strict rule of segregation at the source. Similar to how trash creation can be decreased with improved planning and administration, waste management costs as a whole can be kept under control. Nowadays, a lot of emphasis is placed on organizational structure, training, and motivation. A significant improvement may be achieved by providing all levels of staff at healthcare facilities with proper training and ongoing motivation.

7.2 Components of Biomedical Waste

The various components of BMW are described below and shown in Figure 7.1.

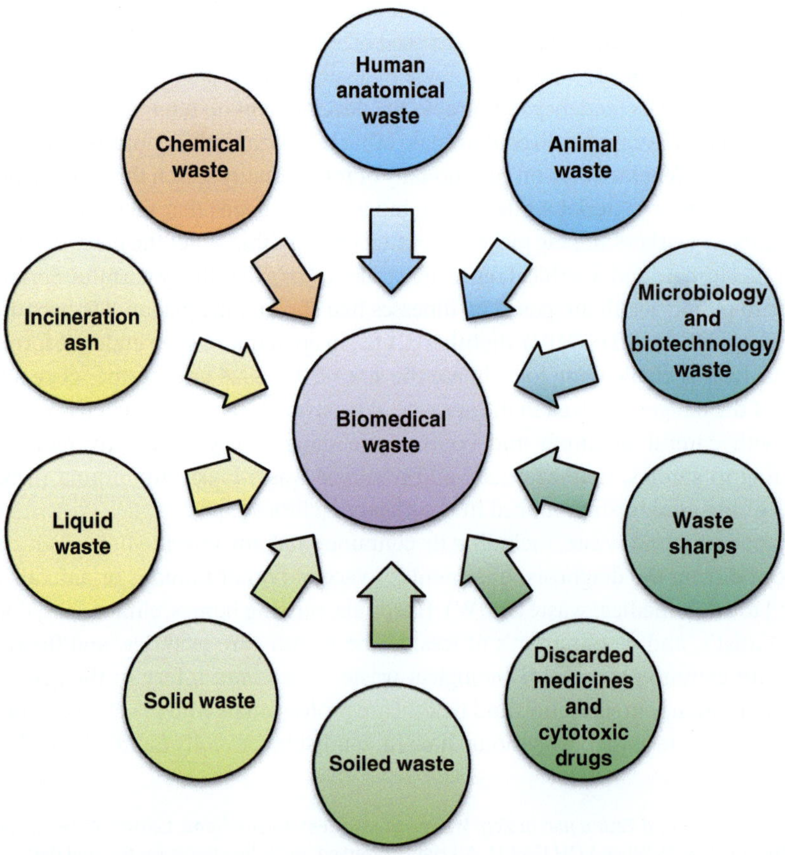

Figure 7.1 Components of biomedical waste.

i. Human anatomical waste (tissues, organs, body parts, etc.).
ii. Animal waste (as above, generated during research/experimentation, from veterinary hospitals, etc.) [6].
iii. Microbiology and biotechnology waste, such as laboratory cultures, microorganisms, human and animal cell cultures, toxins, etc.
iv. Waste sharps, such as hypodermic needles, syringes, scalpels, broken glass, etc.
v. Discarded medicines and cytotoxic drugs.
vi. Soiled waste, such as dressings, bandages, plaster casts, material contaminated with blood, etc.
vii. Solid waste (disposable items like tubes, catheters, etc., excluding sharps).
viii. Liquid waste generated from any of the infected areas.
ix. Incineration ash.
x. Chemical waste.

7.3 Hazardous Waste

Potentially infectious trash has been referred to at various times throughout the years in scientific literature, manuals of best practices, and standards. They include hazardous, red bag, contaminated, regulated, and unregulated medical waste, as well as infectious medical waste. While the names used in regulations are often more explicit, they all refer to the same kind of trash. It accounts for 10% of all garbage, which also includes [7]:

1. Swabbed surfaces and dressings contaminated with blood, pus, and bodily fluids.
2. Lab trash, such as infectious agent culture stocks kept in laboratories.
3. Materials that might be contaminated: placenta, removed organs, and malignancies.
4. Animals utilized in research and diagnostic tests that may be contaminated.
5. Sharp objects, such as needles, syringes, and blades.
6. Blood and its derivatives.

7.4 Management of BMW and Municipal Garbage in Conjunction

Now, there is misunderstanding and laxness on both sides due to the municipal body's unclear position in the management of BMW. As the majority of healthcare facilities are situated in a municipality, the municipal system naturally plays a key role in how they manage their waste. The local authority is also in charge of maintaining public health across the entire municipal jurisdiction. Hence, a clear agreement on the division of responsibility related to this problem between the health care facilities and the town is required. According to studies, about one-third of all hospital trash is neither harmful nor infectious and may even be handled by the municipal waste management system, such as garbage produced by the hospital's kitchen or garden, office, or packing materials from the shop, etc. These tight segregation procedures would reduce the volume and expense of managing hazardous and diseased BMW (collection, transportation, treatment, and disposal). There would need to be shared treatment and disposal facilities under the ownership, oversight, and direction of the local authority since it would not be feasible for every healthcare facility to have its own fully functional system for treating and disposing of biomedical waste [8].

7.5 Potentially Toxic Waste

There are types of potentially toxic waste that are classified as mentioned in Figure 7.2.

7.6 Operation and Maintenance

The operation and upkeep of the biomedical waste management system are of utmost significance once it is set up. Regardless of how large or small the institution is, the management should provide written directives outlining the organization's policy and the choices made that must be followed to all the departments responsible for producing or managing trash. With the assistance of consultants and specialists, charts and timetables should be created and posted in key locations. For the prompt evacuation of garbage, coordination between the local government and the shared treatment and disposal facility is crucial. Regarding the positioning of the waste components, their containers and color coding, the disposal timetable, etc., there should not be any ambiguity [10]. It is crucial to monitor the whole procedure, whether it takes place within the hospital or elsewhere. Both monitoring schedules must be created. In the event that the assigned tasks are onerous or problematic, a backup plan must be created.

Radioactive waste

- This kind of waste comprises waste that has been polluted with radionuclides. This waste might be the result of *in vitro* testing done on human fluids and tissue.

Chemical waste

- It consists of base metal waste, X-ray processing solutions, monomers, and related chemicals (dental amalgam in extracted teeth).

Pharmaceutical waste

- This category comprises used antibiotics, analgesics, sedatives, and anesthetics [21].

Figure 7.2 Potentially toxic waste. *Source*: Adapted from [9].

7.7 Biomedical Wastes Are Divided into Various Categories

Many techniques have been created to make BMW both ecologically safe and visually pleasing, but none of them are appropriate for our situation. The "Biomedical Waste (Management and Handling) Regulations, 1998" specifically listed the 10 main kinds of waste created in healthcare facilities, along with the suggested treatment and disposal methods, as follows: Singh et al. officially listed the 10 main kinds of waste created in healthcare facilities, along with the suggested treatment and disposal methods, as follows [11]:

Category 1 (Human Anatomical Waste)
It consists of human tissues, organs, and body parts and is processed and disposed of by deep burial and incineration.

Category 2 (Animal Waste)
It consists of all animal tissues, organs, body parts, corpses, bleeding parts, etc., that are produced by various health sectors and are handled before being incinerated or buried.

Category 3 (Microbiology and Biotechnology Waste)
Wastes from laboratory cultures, stocks, or specimens of microorganisms employed in research fall under Category 3 (Microbiology and Biotechnology Waste).

Treated locally by cremation, microwaving, and autoclaving.

Category 4 (Waste Sharps)
Needles, syringes, scalpels, blades, glass, and other sharp objects are included in Category 4 (Waste Sharps), and they are handled and disposed of using a chemical disinfectant.

Category 5 (Discarded Medicines and Cytotoxic Drugs)
Consists of expired, tainted, and abandoned medications that are not being used anymore and are handled by incineration, destruction, and drug disposal in landfills.

Category 6 (Solid Waste)
Includes items like dressings, dirty plaster casts, lines, and blood-contaminated cotton that have been treated and handled using the incineration, autoclaving, and microwave methods.

Category 7 (Solid Waste)
Plastic tubing, catheters, intravenous sets, and other solid waste are included in Category 7 and are subjected to chemical disinfection treatment.

Category 8 (Liquid Waste)
Wastes produced by laboratory operations, as well as washing, cleaning, and disinfecting processes, are treated chemically for disinfection before being disposed of into drains.

Category 9 (Incineration Ash)
All biomedical waste that is incinerated falls under Category 9, which is processed and disposed of in municipal landfills.

Category 10 (Chemical Waste)
Chemicals employed in the creation of biological substances are included in this category, as are those that are treated with chemicals and released into drains.

7.8 Different Treatment Technologies for Hospital Wastewater

Hospitals produce a significant amount of wastewater, which is hazardous to the environment and therefore to human health. The effluent produced by hospital departments, including the pathological laboratory, infectious wards, and staff housing, is very complicated since it includes pathogenic bacteria, medication residues, chemicals, biological tissues, cultures, and chemical poisons [12, 13]. The untreated hospital effluent poses a number of potential health risks, including microbial resistance, persistence, DNA damage, toxic effects, ecotoxicity, bioaccumulation, and more [14, 15]. As hospital wastewater has been discovered to negatively impact the aquatic environment, managing hospital wastewater is a challenge for health and environmental authorities on a global scale. Hospital effluents are often blended with household wastewater in several countries and processed directly in municipal wastewater treatment plants (WWTPs) [16]. According to earlier research, hospital effluent includes complex contaminants that must be cleaned before being disposed of in the sewage system to avoid harming aquatic environments and causing an epidemic in the city [17, 18]. Leaving the hospital waste untreated might result in radioactive contamination, cholera and diarrhea epidemics, as well as other major infectious illnesses [19]. As effluent is discharged with pathogenic microorganisms, including pharmaceuticals (drug residues eliminated by urine), human metabolites (traces of blood and body fluids), and industrial chemicals (disinfectants, heavy metals, toxic chemicals, and radioactive elements), as well as posing risks to human health and the environment, hospitals typically require an ample amount of water and almost equal amounts of wastewater. Moreover, the bactericides and disinfectants found in wastewater have the ability to stop the biological processes necessary for wastewater treatment [20]. Multispecialty hospitals (MSHs) treat a wide range of disorders by combining many fields of surgery and medicine under one roof. The effluent produced by these MSH varies in type and properties depending on the specific medical treatment branches. The medications that are utilized are combined together throughout various medical procedures, which further complicates the composition of the effluent. In comparison to a typical hospital setting, the medication residue is highly complicated [21]. Hospitals often employ a variety of harmful compounds, including pharmaceutical waste, cytotoxic medications, disinfectants, active ingredients, pigments, dyes, and chemical reagents [22]. These harmful substances found in wastewater impact living organisms and impair the sewage treatment process. While the hospital effluent is more diluted, there is still a good chance that it might upset the balance of life in the aquatic environment. Prior to being disposed of in the municipal sewage system, hospital effluent must undergo proper treatment in order to safeguard the aquatic system from hazardous chemicals emitted by hospitals [23]. Also, several researches have concentrated on enhanced tertiary treatment to outperform conventional treatment to obtain greater levels of contamination removal of chemicals, pharmaceutical residues, pathogens, etc. [24, 25]. The pharmaceutical industry is thriving in India as a result of the country's expanding economy, and antibiotic output and consumption have both expanded significantly [26]. As a consequence, more antibiotics are being released into the environment. Current research has focused on the detrimental consequences of antibiotics released into the soil and aquatic environment [27]. These antibiotics come from a variety of places, including hospitals, municipal sewage systems, home waste, etc. [28], and their concentration in an ecosystem varies from one nation to another [29]. In addition to these, there are several additional sources of antibiotic discharge, such as the agricultural and pharmaceutical sectors [30]. Antibiotics are used to treat a variety of bacterial and fungal infections in both people and animals [31]. In many agricultural businesses, these antibiotics are also employed as growth boosters and preventative medications [32]. The rates of death and morbidity have decreased as a consequence of the overuse of antibiotics. In a study, antibiotics were even found in the supply of treated drinking water [33]. The emergence of antibiotic-resistant genes, even at low dosages, is the most worrying development of all [34]. Thus, they might be seen as a new hazard with long-term consequences [33]. Most current studies are primarily concerned with describing effluent sources, excluding their influence on treatment effectiveness. Many antibiotic kinds are released into the environment, and their priority is determined by a number of factors, including size, physio-chemical characteristics, toxicity, degradability, and treatment resistance. Using cutting-edge analytical tools, more than 300 medications were studied, and the ranking was done taking into account the growing influence on the environment, the emergence of antibiotic resistance genes, and reproductive problems [35]. The removal of pharmaceuticals from wastewater has been studied using a variety of physical, chemical, biological, and advanced treatments, such as the activated sludge process, membrane bioreactors (MBRs), nanofiltration, reverse osmosis, ozonation, UV/H_2O_2, constructed wetlands (CWs) [36], homogeneous and heterogeneous Fenton processes, and MBRs coupled with ozonation [37]. Among others, using sponge-MBR and ozonation together might be a suitable approach for removing pharmaceuticals from wastewater [38]. Also, several MBR technologies minimize antibiotics and organic compounds and eliminate a variety of microbes from hospital wastewater to improve public health and environmental protection [38]. Additionally, the combination of these seven technologies with advanced oxidation processes (AOPs) was little acknowledged in the literature. Thus, thorough research into the removal of pharmaceutical chemicals from wastewater using different treatment techniques in conjunction with AOPs is necessary for full treatment.

Hospital effluents are the source of a complicated, high proportion of micropollutants, but no specific removal method has been researched up to this point. These effluents' intricate features make them resistant to the unit processes used in hospital WWTPs. In other words, standard treatment methods are unable to entirely eliminate such complex pollutants. The world's water supplies might be seriously threatened by pharmaceutical residues that are released into the environment. Environmental experts have recently been increasingly concerned with the effluent produced by healthcare institutions, community health clinics, and hospitals. In this research, three aspects, conventional wastewater characteristics, high-priority medications, and microbiological analyses, were examined in seven MSHs with nonidentical pretreatment. Here, we'll talk about and assess the effectiveness of sophisticated wastewater treatment techniques for removing these three elements from hospital effluents before releasing them into a sewage treatment plant (STP).

Except for the regeneration of microorganisms, which was shown at the tertiary level by these two treatments, MBR and CW efficiently reduced conventional parameters and medicines in secondary and tertiary treatments. The results further imply that although CW showed 100% ofloxacin clearance, MBR showed 100% elimination of ibuprofen, carbamazepine, and frusemide. Moreover, it has been shown that the advanced oxidation processes of ozonation (O_3) and peroxone (O_3-eH_2O_2) may oxidize pharmaceuticals. For AOP, pH, O_3 supply, contact duration, and H_2O_2 concentrations are the operationally important parameters. Based on pharmaceutical reduction efficiencies, the traditional O_3 alone is more effective than the combined O_3 and eH_2O_2 therapy. So, it is confirmed that among the seven methods evaluated for the pretreatment of hospital wastewater, MBR in combination with the ozone process was the most effective [38].

7.9 Incineration

This is a high-temperature thermal process that uses controlled combustion of the waste to turn it into inert material and gases. Incinerators may run on electricity, oil, or a mix of the two. Hospital trash is often burned in three different kinds of incinerators: controlled air, rotary kilns, and multiple hearths. To guarantee optimum combustion, all varieties are capable of having main and secondary combustion chambers. They are lined with refractory.

The main chamber of the multiple-hearth incinerator is used for solid-phase burning, while the secondary chamber is used for gas-phase combustion. Since there is extra air in both chambers, they are known as surplus air incinerators. The rotating kiln is a cylindrical, refractory-lined shell that is slightly tipped to make it easier for the waste to be mixed and moved about within. It is equipped with a ventilation system. The main solid-phase chamber is the kiln, and the secondary chamber for gaseous combustion comes next. In the third kind, an overflow air chamber is used after the first chamber, which is run at low air levels. The main chamber's low oxygen content helps to better regulate the amount of particulate matter in the flue gas.

In short, the main chamber features pyrolytic conditions with an approximate temperature range of 800 (+/−) 50 °C. At around 1,050 °C (+/−50 °C), the secondary chamber functions in an environment with surplus air (Schedule V of the

Rules). In the first chamber, the volatiles are released, but in the second, they are destroyed. Certain types have an eductor mechanism that keeps the system under negative pressure and aids in better management of the exhaust gases. A minimum height of 30 m above ground level is required for the chimney. It is not advisable to install incinerators in a crowded region.

For human anatomical waste, animal waste, cytotoxic pharmaceuticals, abandoned medicines, and dirty trash, incineration has been advised under the Biomedical Waste (Management and Handling) Regulations.

7.10 Autoclave Treatment

This is a pressure-assisted steam sterilization procedure. It is a low-heat procedure in which the waste material is brought into close contact with steam for long enough to disinfect it. They come in three different varieties as well: gravity type, prevacuum type, and retort type. In the first kind (the gravity type), just gravity is used to expel the air. At a steam pressure of 15 psi and a temperature of 121 °C, the system is operational for 90–120 minutes. In order to shorten the prevacuum autoclave cycle to 30–60 minutes, air is removed using vacuum pumps. It runs at about 132 °C. Autoclaves of the retort type are designed to handle significantly larger quantities and run at much higher steam pressure and temperature. Waste from microbiology and biotechnology, sharps waste, and dirty and solid waste have all been advised to undergo autoclave treatment. This technique makes certain types of biomedical waste (included in the guidelines) harmless and unrecognizable so that the treated residue may be disposed of in landfills. A prevacuum autoclave has been built at Delhi's Sanjay Gandhi Memorial Hospital.

7.11 Hydroclave Treatment

A piece of cutting-edge machinery for steam sterilization is the hydroclave (like an autoclave). It is a double-walled container, and to heat the inner chamber holding the waste, steam is fed into the outer jacket. The waste's moisture evaporates as steam, creating the necessary steam pressure (35–36 psi). The garbage is continually tumbled against the heated wall within the chamber by sturdy paddles that are gently rotated by a powerful shaft, mixing and fragmenting the waste as they do so. More steam is injected if there is not enough moisture present. The system runs at 132 °C and 36 psi steam pressure for 20 minutes of sterilization. The duration of a cycle, which comprises start-up, heat-up, sterilization, venting and depressurization, and dehydration, is around 50 minutes. Before disposal, the material might be further shredded. Up to 85 and 70%, respectively, of the volume and weight reductions are anticipated. Together with the waste sharps, the hydroclave may handle the same waste as an autoclave. Sharps are also broken apart. Some advantages of this technique are the lack of hazardous air emissions, the absence of liquid discharges, the lack of chemical requirements, the reduction of waste volume and weight, etc.

In September 1999, the first hydroclave was erected in India at the Tata Memorial Hospital in Mumbai.

7.12 Microwave Treatment

Again, this is a wet thermal disinfection method, but a microwave sterilizes the target material to a great degree by heating it from the inside out, as opposed to other thermal treatment methods that just heat the waste outwardly. First, a shredder is used to process the input material. Pushing the shredded material into a treatment chamber allows high-temperature steam to wet it.

The material is subsequently transported by a screw conveyor under a row of standard microwave generators (typically 4–6 units), which heat the material to 95–100 °C and evenly disinfect it throughout the course of a cycle that lasts 50 minutes. Before being mechanically disposed of into a conventional or general garbage receptacle, a second shredder further shreds the item into unidentifiable bits. If the microwave treatment is carried out properly, the waste may be disposed of in a landfill. In more recent iterations, the process control is automated for smooth and efficient control.

Some advantages of microwave technology include the absence of harmful air emissions, the absence of liquid discharges, the lack of chemical requirements, the reduction of waste volume due to shredding and moisture loss, and operator safety (due to the automatic hoisting arrangement for the waste bins into the hopper so that manual contact with the waste bags is not necessary). Yet, the cost of investment is now significant.

The following categories are allowed to be microwaved: nos. 3 (waste from microbiology and biotechnology), 4 (trash from sharps), 6, and 7 (solid waste).

7.13 Chemical-based Treatment

Several of the chemicals used to dispose of BMWs are still being developed. Chemical-based technology treats a variety of wastes, including soft trash, liquid waste, human waste, sharps waste, and laboratory waste (gauze, bandages, and gowns). Mercury, semi-volatile organic chemicals, and radioactive waste should not be handled using chemical-based techniques. To prevent aerosol formation during shredding, the chemical-based technique needs closed systems or is operated under negative pressure, and the expelled air must be filtered through high-efficiency particulate air (HEPA). BMW must be destroyed via chemically based technologies. Chemical-based technology has many benefits, including a completely automated process, convenience of use, ease of dumping liquid effluent into sewers, and the absence of combustion by-products. The drawbacks include the development of harmful by-products from the widespread use of chlorine and hypochlorite, chemical risks, and often disagreeable odors. Chlorine-based and nonchlorine-based systems make up chemical-based technologies. Chlorine dioxide or sodium hypochlorite is used in the chlorine-based system. One of the earliest chemical disinfectants used to cure BMW was sodium hypochlorite. Recently, it has been shown that the usage of sodium hypochlorite causes the production of toxins such as dioxins, haloacetic acid, and chlorinated aromatic compounds. As chlorine dioxide is unstable, it is produced and utilized locally. It is an effective biocide. It does not react with ammonium or alcohol, breaks down to create salt, and produces less pollutants when used. Nonchlorine-based approaches for treating BMW employ either gas, liquid, or dry chemicals. Numerous of these systems have entered the market, but just a handful are included here. The SteriEcocycle 10-4 employs a portable chamber to collect trash, into which a vial of a decontaminant with a peracetic acid base is introduced. After 10–12 minutes, peracetic acid cleans the waste, and a HEPA filter prevents aerosolized microorganisms from escaping. The garbage is dumped as usual into the sewage along with the liquid effluent. Alkaline hydrolysis at high temperature was used in waste reduction (WR) technology to transform human and microbiological waste into a neutral aqueous solution [39]. Body fluids, human or tissue waste, and biodegradable bags are utilized in it. It can also manage chemotherapeutic waste. The alkali also breaks down dangerous compounds like glutaraldehyde and formaldehyde, as well as fixatives in tissues. For biodegradable trash, such as bones and teeth, this automated system features a steam-jacketed stainless steel container with a retainer. Ozone is used by Lynntech in another method for cleaning BMWs. Strong oxidant ozone kills bacteria and transforms them into molecular oxygen [40]. Ionizing radiation damages DNA and further damages infectious particle proteins and enzymes by generating free radicals [40]. Ionizing radiation is created by the electron beam technology in the form of a high-energy electron beam that is accelerated quickly toward the target. The garbage that has to be processed includes sharps, human waste, laboratory waste, soft waste (gauze, bandages, and gowns), and contagious waste. Mercury, radioactive waste, and volatile or semi-volatile organic chemicals should not be processed in electron beam systems. This technology has a number of benefits, including the absence of harmful emissions and liquid effluent, the absence of ionizing radiation after machine shutdown, complete automation, and minimal operating costs. The drawbacks include the necessity for shredders or grinders, protection from radiation exposure, a concrete barrier several feet thick surrounding the system, the removal of ozone gas, and no reduction in waste volume. A developing technology known as "Bioconverter" (9 Biomedical Disposal, Inc.) is one biological way for disposing of BMW [40]. It decontaminates medical waste using an enzyme solution, and the resultant sludge is pushed through an extruder to separate water for sewage disposal while the solid waste is transferred to a landfill. Biodegradable plastics are an additional means of BMW disposal that is environmentally friendly. A large number of biomedical implants made of biodegradable polymers are biologically degraded by microbial extracellular enzymes. These microorganisms use these biodegradable polymers as a substrate when they are starved or when there is no other available substrate. In order to create biodegradable polymers on a large scale economically, further research must be conducted [40, 41].

7.14 Waste Survey

It is a crucial part of the waste management process. A survey aids in determining the kind and volume of garbage produced [42, 43].

The waste survey is useful in that it can:

- Distinguish between different forms of garbage.
- Please list the garbage produced.
- Identify the waste types produced at each site of production.
- Establish the hospital's level of generation and disinfection.
- To determine the sort of disposal that was done.

7.15 Segregation of Waste

- Segregation is the fundamental separation of several waste types that are produced on a regular basis, lowering the hazards as well as the expense of handling and disposal.
- The most crucial phase of managing BMW is segregation. Only effective segmentation can guarantee the efficient handling of BMW.

At the time of creation, various waste types are put in distinct containers or bags with codes. As a result, costs associated with infectious waste treatment and disposal are reduced. As a result, costs associated with infectious waste treatment and disposal are reduced. Separation also slows the transmission of disease and lowers the risk of contaminating other medical professionals. (Biomedical Waste Management, Govt. of India)

7.15.1 BMW Collection and Segregation (Biomedical Waste Management, Govt. of India)

The best methods for managing trash begin at the point of creation, when garbage is formed, as is the case with all waste. As each category of medical waste needs to be segregated from the other sorts, one must first be aware of precisely what to collect. In actuality, this implies using the appropriate containers. They are often color-coded to aid in the separation of BMW as mentioned in Figure 7.3 [44].

- **Red Containers:** Collection of sharps waste (e.g., needles, blades, and razors).
- **Red Containers Marked with the Biohazard Symbol:** Collection of infectious waste (e.g., blood, contaminated equipment, IV tubing).
- **Yellow Containers:** Collection of trace chemotherapeutic waste (e.g., empty vials, gloves, gowns).
- **Black Containers:** Collection of hazardous garbage (e.g., hazardous medications, P-listed drugs, bulk chemotherapy).
- **Blue Containers:** Collection of pharmaceutical waste (e.g., pills, injectables, antibiotics).

Figure 7.3 Biomedical waste collection and segregation using color-coded containers.

- **Yellow-protected Containers Marked with the Biohazard Symbol:** Collection of radioactive waste with a radioactive symbol (e.g., lab research liquids, anything contaminated by radiotherapy).

7.16 Storage and Accumulation of BMW

Between the site of trash creation and the place of waste treatment and disposal, waste is accumulated and stored in the stages and spaces in between. While the term "accumulation" refers to the short-term keeping of tiny amounts of garbage close to the source of creation, the term "storage" refers to longer holding times and larger waste volumes. Storage spaces are often situated close to the waste treatment facilities. Any storage of garbage off-site is likewise regarded as storage [45].

For the collection of garbage, a variety of containers are employed.

- The containers or bins need to be placed such that 100% of the material is collected.
- Sharps must be stored in puncture-proof containers to protect workers handling them from infections and accidents.
- The biological waste is subsequently placed in the proper location when collection is complete.
- Separated garbage from different categories must be gathered in separate containers or bins.

- Each container or bin should be properly labelled to indicate the ward or room where it is stored.
- Each container or bin should be properly labelled to indicate the ward or room where it is stored. Also, a warning sign should be placed at the storage location to draw attention to it. This labeling is necessary since it could be necessary to track the trash back to its source.

7.17 Transportation of Waste

Carts and containers that are not used for any other purpose may be used to transport biomedical waste. Every day, the carts must be cleaned. The name and address of the transporter should be clearly displayed on all off-site carrying vehicles. The biohazard sign needs to be colored. It is important to have a reliable mechanism in place to keep the weight secure while being transported. With rounded edges, such a vehicle should be easy to clean. When returning any throwaway plastic to the seller, it should be shredded. The maximum amount of time that unprocessed BMW is kept is 48 hours [46].

7.18 Treatment of Waste

The term "treatment" refers to a process that alters trash in some manner before it reaches its final resting place. The waste itself must be cleaned up or decontaminated at the source in order to stop harmful organisms from growing there. After receiving such care, the remains may be handled, transported, and stored securely [47].

- Syringe cutters and needle destroyers should be used to shred syringe nozzles and needles.
- Broken glass, scalpel blades, and lancets need to be stored in separate containers with bleach, moved to plastic or cardboard boxes, sealed to avoid spills, and delivered to incubators.
- Glassware has to be cleaned, sterilized, and disinfected.
- Culture plates that may contain a culture must be autoclaved; the media are then put in appropriate bags and thrown away.

After sterilization, the plates may be used again.

Swabs should be chemically sterilized and then burned. Gloves should be torn, shredded, or otherwise damaged before disposal. They could be thrown out if they merely contain a little amount of blood that does not drop.

- The policy for such products should be to mutilate them first before disposal after being immersed in newly made 1% sodium hypochlorite for 30 minutes to 1 hour.
- Under no circumstances may heat be used for the disposal of amalgam. It will lead to the volatilization and release of mercury into the environment. As a

result, amalgam-filled teeth should be treated by submerging them for 30 minutes in high-level disinfectants such as glutaraldehyde.
- The laboratory generates either pathological or chemical liquid waste. Reagents should be used to diffuse noninfectious trash.
- Chemical decontaminators should be used to treat liquid-contaminated trash before defusing it.

7.19 Reduction of Waste

Regular solid and liquid garbage may be disposed of without treatment, but almost all infectious waste has to be treated beforehand. Disposing of infectious garbage might cost 10 times as much as disposing of regular waste. Any strategies that lower the amount of infectious waste generated will also lower the cost of infectious waste disposal [48].

7.20 Healthcare Plastics Management

Surgical gloves, catheters, disposable syringes, blood, urine, and bottle bags are a few examples of plastics used in healthcare. Sperm count decline, genital anomalies, and an increase in incidence have all been linked to plastic. Surgical gloves, catheters, disposable syringes, blood, urine, and bottle bags are a few examples of plastics used in healthcare. Plastic has been linked to an increase in breast cancer cases, genital abnormalities, and a decline in sperm count. Burning plastic releases dioxin and furan, two known carcinogens. Plastic's inability to biodegrade makes it a serious environmental and health hazard. Environmentally responsible methods of disposal for plastic trash include recycling or long-term landfilling. Before giving any discarded plastic to the seller, it should be shredded. The current need is to create biodegradable, environmentally friendly polymers. It is also essential to reduce the amount of plastic garbage produced [49].

The handling of biological waste was controlled by many laws. In accordance with Sections 6, 8, and 25 of the Environment (Protection) Act of 1986, the Central Government announced these regulations on July 20, 1998. Prior to that, on October 16, 1997, the draft regulations were published for public discussion and recommendations, which included a 60-day comment period. Before setting the guidelines in stone, these comments were taken into account. The Biomedical Waste (Management and Handling) Regulations, 1998, issued by the Ministry of Environment and Forests (MoEF), Government of India, and published on July 20, 1998, set standardized standards and norms for the whole country. The "occupier" of an institution that generates biomedical waste (such as a hospital, nursing home, clinic, dispensary, veterinary institution, animal house, pathological laboratory, blood bank, etc.) is required by this rule to take the necessary precautions to ensure that such waste is handled in a manner that protects both human health and the environment (Govt. of India, Biomedical Waste Management Rules).

In India, there are a number of laws governing the handling of BMW, including the following:

- The 1974 Water Act (prevention and control of pollution).
- The 1981 Air Act (prevention and control of pollution).
- The 1986 Environment Act (Protection).
- The 1998 Hazardous Waste Rules (management and handling).
- The 1998 BMW Regulations (management and handling).
- The 2000 Municipal Solid Waste Rules (management and handling).
- Modifications to the BMW Rules, 2000 and 2003 (management and handling).
- The 2011 BMW Regulations (Management and Handling).
- It should be remembered that anybody may contact the appropriate authorities to report any suspected carelessness in the management and handling of biomedical waste [50].

7.21 Environmental Issues Related to the Treatment of BMW

The biggest environmental issues with incorrect BMW management and disposal include the following [47]:

- The transmission of illness and infection by insects, flies, and other vectors that harm both the occupants of the building and the surrounding community
- Infection transmission via contact with or damage to medical, nonmedical, and rag-pickers and sweepers, particularly from sharps (needles, blades, etc.)
- The improper recycling of disposable objects like hypodermic needles, tubes, blades, bottles, etc., might cause an illness to spread
- Response brought on by the usage of expired medications
- Toxic emissions from incinerators that are broken or ineffective. disposal of incinerator residues and ash without consideration

7.22 Tie-up of Healthcare Organizations

First, a clear choice must be made on the model to be used, such as Build, Own, Operate, and Maintain (BOOOM), Build, Own, Operate, and Transfer (BOT), etc. The following factors need to be taken into account for a long-term partnership and efficient operation of the shared treatment and disposal facility [51]:

1. **Legal Considerations:** Creation and execution of a legal contract between the parties, i.e., the healthcare facilities Approval from the appropriate authorities, in accordance with the most recent government announcement, such as the "prescribed authority" (Pollution Control Board or other authorities notified by the state government)

2. **Financial Considerations:** Whole system costing (capital cost, depreciation, O&M cost, interest, debt payment, etc.). calculating a capital contribution that is proportionate. developing a cost-sharing system. For instance, a system of monthly invoicing and advance payments for all recurrent charges
3. **Management-related Factors:** Establishing an action committee or group to help with facility management and oversight on a daily basis

7.23 Alternative Technology for BMW Disposal

The numerous innovative techniques for disposing of BMWs are divided into four categories: thermal, chemical, irradiative, and biological processes. The majority of these are currently being studied. Low, medium, and high thermal processes are classified into three categories. (i) Microwaves and autoclaves are examples of low-heat technologies that operate between 93 and 177 °C [33]. Steam is used in autoclaves as a sterilization technique. According to how air pockets are removed, autoclaves are divided into two categories: gravity/downward displacement and prevacuum/high vacuum. They are superior autoclaves because air evacuation is more efficient in those with prevacuum or repeated vacuum cycles. If garbage is to be rendered unrecognizable or the amount of waste is to be reduced, a shredder or grinder should be employed. Before being released into the environment, the evacuated air is cleaned by passing through a HEPA filter since it could contain diseases. The autoclave is used to sterilize all infectious waste, including cultures, human waste, laboratory trash, soft waste (gauze, bandages, and gowns), sharps, and medical equipment. Chemicals and hazardous trash cannot be autoclaved because they emit harmful fumes. Moreover, bulky garbage, mattresses, and containers that can withstand heat cannot be disinfected in an autoclave. The steam and moist heat produced by the microwave are used to disinfect. Microwave sterilization is used to kill all infectious waste, including human waste, laboratory waste, soft waste (gauze, bandages, and gowns), and sharps [33]. Mercury, semi-volatile organic chemicals, radioactive waste, and volatile organic compounds should not be microwaved. Little emissions are a benefit of using a microwave for BMW disposal. Its drawbacks include the need for a significant startup cost, an odor issue around the equipment, and the potential for microwave energy leakage. (ii) Reverse polymerization and thermal depolymerization are examples of medium-heat technologies that work between 177 and 540 °C. In order to break down the organic stuff, high-energy microwaves are applied to the BMW in a nitrogen environment. The waste's internal energy increases as the microwave radiation is absorbed, and chemical disintegration occurs at the molecular level. In order to prevent combustion, nitrogen creates an atmosphere devoid of oxygen. The garbage is then mutilated using shredders [33]. (iii) High-heat technologies, such as pyrolysis-oxidation, plasma pyrolysis, induction-based pyrolysis, and laser-based pyrolysis, operate between 540 and 8,300 °C. In pyrolysis oxidation, the organic solid and liquid waste evaporate within the pyrolysis chamber at a high temperature (about 594 °C), leaving behind inert ash and glass and metal shards. The next phase involves burning the vapors at a temperature between 982 and 1,093 °C in a chamber, after which clean exhaust steam is emitted. Plasma energy is produced via plasma pyrolysis using plasma torches. The ionized gas may carry electricity in plasma form, but because of its high resistance, the electricity is converted to heat energy. Carbon black, vitrified glass aggregates, and metallic residues are among the by-products produced. Plasma-based technology may eliminate a range of trash, including low-level radioactive waste, infectious waste, sharps, plastics, dialysis waste, hazardous waste, and chemotherapeutic waste (one exemption is mercury, which plasma systems do not handle). This technology has several benefits, including low emission rates, inert and sterile waste residue that is environmentally benign, and reductions in volume (by 95%) and mass (by 80–90%). Its drawbacks include expensive startup costs, high operating costs, high electricity use, a plasma torch with a short lifetime, and the potential for dioxin production in an improperly built system [52].

7.24 Potential Adsorbent for a Variety of Organic Dyes and Simulated Hospital Effluents Using Nanocomposite Materials

Hospital effluents and industrial textile wastewater may both include a variety of colors, including Congo red, methyl red, rhodamine B, methyl orange, methylene blue, and crystal violet (Figure 7.1). Such effluents are known to include organic dyes, which are a key cause of a variety of harmful effects on aquatic ecosystems [46]. Past studies have shown that they might have persistent teratogenic and carcinogenic effects on people [49, 50]. However, a lot of them show great stability and bio-resistance to photocatalysts and oxidants, which makes it difficult to use physically and biologically altered technologies. Adsorption techniques may be the best options for water filtration in terms of treatment efficacy and cost-effectiveness [53].

To separate the solid adsorbent from the aqueous solution during the post-adsorption process, strong centrifugation must be used. Porous carbons with surface functional groups were of particular interest for removing organic dyes [53]. As a result, magnetically loaded carbon sorbents (usually Fe_3O_4@C) make the most of both their favorable properties and ease of separation, improving both their removal efficiency and their ability to adsorb a variety of contaminants. For instance, Zhou et al. (2017) found that Fe_3O_4@C nanospheres functionalized with active chemical groups efficiently treated Cr(VI) with a high adsorption capacity (331.8 mg/g) [54]. The simultaneous removal of Pb(II) (92%) and Rhodamine B (98%) from TiO_2-doped Fe_3O_4@C nanocomposites is particularly notable [53]. Fe_3O_4 has shown several favorable outcomes for its specialized use in the treatment of dyes. As a result, $FeCl_3$ was used as an iron source to modify KOH-activated semi-coke by Gao et al. (2016) such that a number of dyes could be adsorbed onto it [18]. Cho et al. (2016) demonstrated excellent performance for the adsorption of methylene blue (52.6 mg/g) by Fe_3O_4 at C during the pyrolysis of a sugar combination containing Fe_3O_4 at a temperature of 500 °C [14]. Cazetta et al. (2016) developed a unique magnetic nanocomposite from coconut shells and $FeCl_3$ via many steps (impregnation, pyrolysis, and purification) for improved removal of acid yellow 6 and sunset yellow food coloring [11]. According to Altintig et al. (2018), magnetically loaded and unloaded carbons have significantly different capacities for absorbing the dye malachite green (311.40 and 103.64 mg/g, respectively) [1]. Nevertheless, nearly all manufacturing calls for a multistep process that involves combining carbon precursors and Fe_3O_4 nanoparticles (NPs). To the best of our knowledge, relatively few investigations have focused on straightforward manipulation of Fe_3O_4@C NPs and direct production of these materials for wastewater cleanup.

In this case, six organic dyes from aqueous solutions and simulated hospital effluents were treated with Fe_3O_4@C nanocomposite. The Fe_3O_4@C surface has a variety of functional groupings due to its distinguishing characteristics (e.g., OH, COOH). Ultrathin carbon coatings seem to enclose core-shell Fe_3O_4 NPs with a thickness of 4–5 nm and well-distributed Fe_3O_4 NPs (average size: 40 nm). According to the adsorption data, organic dyes' maximal adsorption capacities on Fe_3O_4@C conformed to the following rule: crystal violet (181.6 mg/g), methyl orange (38.03 mg/g), rhodamine B (87.32 mg/g), methylene blue (95.61 mg/g), methyl red (153.1 mg/g), and Congo red (165.3 mg/g). With this nanocomposite, two simulated hospital effluents may also be treated to a degree of 63.6–84.5%. Fe_3O_4 may also be recycled up to five times without significantly losing absorption capacity. Notwithstanding certain restrictions on the use of organic dyes and simulated hospital effluents, positive results allow us to suggest Fe_3O_4@C as a possible adsorbent for the practical domains of wastewater treatment and purification.

7.25 Financial Aspects of Managing BMW

The extra cost needed for appropriate biomedical waste management is not significant when compared to the cost of medical facilities and capital, as well as operation and maintenance. The following issues need careful thought: The Biomedical Waste (Management and Handling) Regulations, 1998, now make correct management of biomedical waste essential in addition to being a social and ethical requirement of the relevant institutions and people. It should typically be feasible to save some money by implementing a more efficient system, improving control over people and materials, and avoiding the expense of accidents and compensation [45].

Mobilization of resources for undertaking capital investments includes loans, grants, and individual contributions. For the purpose of covering O&M expenses and debt payments, it may be investigated to add a little additional charge to the different medical institutions' current service fees. Similar issues must be discussed by the relevant institutions in the case of the shared treatment facility. If the local authority chooses to build the facility using its own funds or a loan (BOT or BOOT model), the user institutions will be required to pay proportionate fees, which include capital as well as O&M costs. Similar problems would still apply if private funds were used to create a BOO model, but the transactions would take place between the individuals involved and the facility. In this situation, the city body may lease out property and set a reasonable leasing fee. Financial institutions provide financing (loans) [46].

7.26 Training and Motivation

The training program seeks to educate the management and provide the medical, paramedical, and support personnel with the practical skills and instructions they need for their individual tasks. A core group of trainers should be created at the same time for ongoing internal training of the auxiliary and sanitation employees. The following types of functionaries and persons must get training and orientation courses in order for the program to succeed. Based on these individuals' backgrounds, skills, and responsibilities, their roles and responsibilities are as follows:

- Municipal authorities
- Policymakers
- Hospital administration, medical directors, deans, and department heads
- Ward boys
- Cleaning/sanitation personnel
- Auxiliary and nursing staff
- Physicians, surgeons, and specialists.

It is necessary to repeat the training program for each category, particularly for the auxiliary workers. The management must determine how long to wait between two programs, taking into account the number of employees and other resources that are available.

From the perspective of a planner, the subject of hospital waste management may be separated into two categories:

i. Internal (hospital/nursing home planning, multiple movement routes, service corridors, hospital waste storage area, potential hospital waste treatment site, accessibility to open space, buffer zone, etc.).

ii. Site selection for external (off-site treatment and disposal, common treatment and disposal facility) facilities is done in accordance with the town's land use plan, transfer route, etc. Although some of these may still apply to already-existing facilities or the goal of expansion or extension, they have been presented with the needs at the time of original planning in mind.

7.27 Conclusion

For medical experts, managing BMW is just as crucial as developing a treatment strategy. All healthcare professionals and support workers at different healthcare facilities should participate in awareness programs to stay up-to-date on the scientific BMW management system's significance and advantages for patients, employees, and the community at large. An update, continuing medical education, and ongoing education should be conducted periodically. Scientific judgment must be utilized in combination with a good legal/regulatory framework for programs in health care waste management to be effective. It is advised that the systems for the safe handling of medical waste be periodically monitored. The occupational health and safety of healthcare workers should take precedence since they are the ones who would be most in danger if waste were improperly handled. Each health care centers/units should adapt and use a method that works best for them in order to enforce the notification of BMW (handling and management) requirements, 1998, strictly and transparently. Every community member has a right to know about the negative impacts of medical waste. So, via mass media, the people must be well informed. Health care centers/units should use a transparent, all-encompassing approach to providing medical services if they want to attain quality in the disposal of medical waste. This should include how they handle their trash in terms of the environment. BMWM should be a collaborative effort with strong legislative support, dedicated government backing, and excellent BMW practices adhered to by both healthcare professionals and HCFs. Our basic right to live in a sanitary and secure environment. Waste separation at the source and WR are the cornerstones of BMWM. In terms of better segregation, transportation, and disposal techniques, the new BMWM 2016 guidelines outperform older regulations in order to reduce environmental pollution and guarantee the safety of the staff, patients, and general public. Also, it is important to promote the adoption of non-PVC medical equipment and the development of innovative, eco-friendly BMW disposal methods. A cleaner and greener environment should be guaranteed by all BMWM participants.

References

1. Altintig, E., Onaran, M., Sarı, A., et al. (2018). Preparation, characterization and evaluation of bio-based magnetic activated carbon for effective adsorption of malachite green from aqueous solution. *Mater Chem Phys.* 220: 313–321.
2. Auvinen, H., Havran, I., Hubau, L., et al. (2017). Removal of pharmaceuticals by a pilot aerated sub-surface flow constructed wetland treating municipal and hospital wastewater. *Ecol Eng.* 100: 157–164.
3. Baluch, S. (2017). Evaluation of waste management practices in one of the leading public sector hospitals at Karachi. *Glob Manag J Acad Corp Stud.* 7 (2): 14–14.
4. Bano, K., Kuddus, M., Zaheer, M.R., et al. (2017). Microbial enzymatic degradation of biodegradable plastics. *Curr Pharm Biotechnol.* 18 (5): 429–440.
5. Bhalla, G.S., Bandyopadhyay, K., and Sahai, K. (2019). Keeping in pace with the new biomedical waste management rules: what we need to know! *Med J Armed Forces India.* 75 (3): 240–245.
6. Bi, J., Huang, X., Wang, J., et al. (2019). Oil-phase cyclic magnetic adsorption to synthesize Fe_3O_4@C@ TiO_2-nanotube composites for simultaneous removal of Pb (II) and rhodamine B. *Chem Eng J.* 366: 50–61.

7 Binh, V.N., Dang, N., Anh, N.T.K., and Thai, P.K. (2018). Antibiotics in the aquatic environment of Vietnam: sources, concentrations, risk and control strategy. *Chemosphere*. 197: 438–450.

8 Block, S.S. (ed.) (2001). *Disinfection, Sterilization, and Preservation*. Lippincott Williams & Wilkins.

9 Boxall, A.B., Rudd, M.A., Brooks, B.W., et al. (2012). Pharmaceuticals and personal care products in the environment: what are the big questions? *Environ Health Perspect*. 120 (9): 1221–1229.

10 Bui, X.T., Chen, S.S., Nguyen, P.D., et al. (2019). Hospital wastewater treatment by sponge membrane bioreactor coupled with ozonation process. *Chemosphere*. 230: 377–383.

11 Cazetta, A.L., Pezoti, O., Bedin, K.C., et al. (2016). Magnetic activated carbon derived from biomass waste by concurrent synthesis: efficient adsorbent for toxic dyes. *ACS Sustain Chem Eng*. 4 (3): 1058–1068.

12 Chandra, H., Jamaluddin, K., Masih, L., and Agnihotri, K. (2006). Cost-benefit analysis/containment in biomedical waste management: model for implementation. *J Financ Manag Anal*. 19 (2): 110–113.

13 Changotra, R., Rajput, H., Guin, J.P., et al. (2020). Techno-economical evaluation of coupling ionizing radiation and biological treatment process for the remediation of real pharmaceutical wastewater. *J Clean Prod*. 242: 118544.

14 Cho, D.W., Lee, J., Ok, Y.S., et al. (2016). Fabrication of a novel magnetic carbon nanocomposite adsorbent via pyrolysis of sugar. *Chemosphere*. 163: 305–312.

15 Czekalski, N., Gascón Díez, E., and Bürgmann, H. (2014). Wastewater as a point source of antibiotic-resistance genes in the sediment of a freshwater lake. *ISME J*. 8 (7): 1381–1390.

16 Deepak, A., Sharma, V., and Kumar, D. (2022). Life cycle assessment of biomedical waste management for reduced environmental impacts. *J Clean Prod*. 349: 131376.

17 Forgacs, E., Cserháti, T., and Oros, G. (2004). Removal of synthetic dyes from wastewaters: a review. *Environ Int*. 30 (7): 953–971.

18 Gao, X., Zhang, Y., Dai, Y., and Fu, F. (2016). High-performance magnetic carbon materials in dye removal from aqueous solutions. *J Solid State Chem*. 239: 265–273.

19 Gautam, A.K., Kumar, S., and Sabumon, P.C. (2007). Preliminary study of physico-chemical treatment options for hospital wastewater. *J Environ Manag*. 83 (3): 298–306.

20 Gholipour, S., Nikaeen, M., Farhadkhani, M., and Nikmanesh, B. (2020). Survey of *Listeria monocytogenes* contamination of various environmental samples and associated health risks. *Food Control*. 108: 106843.

21 Hegde, V., Kulkarni, R.D., and Ajantha, G.S. (2007). Biomedical waste management. *J Oral Maxillofac Pathol*. 11 (1): 5.

22 Khan, N.A., Ahmed, S., Vambol, S., et al. (2019). Field hospital wastewater treatment scenario. *Ecol Quest*. 30: 57–69.

23 Klaus, K. (2009). Antibiotics in the aquatic environment–a review–Part II. *Chemosphere*. 75 (4): 435–441.

24 Klinck, J.S., de Solla, S.R., Gilroy, È.A., et al. (2016). Bioaccumulation of pharmaceuticals and personal care products in the unionid mussel lasmigonacostata in a river receiving wastewater effluent. *Chemosphere.* 146: 486–496.

25 Kovalova, L., Siegrist, H., Singer, H., et al. (2012). Hospital wastewater treatment by membrane bioreactor: performance and efficiency for organic micropollutant elimination. *Environ Sci Technol.* 46 (3): 1536–1545.

26 Kumar, A., Duggal, S., Gur, R., et al. (2015). Safe transportation of biomedical waste in a health care institution. *Indian J Med Microbiol.* 33 (3): 383–386.

27 Kummerer, K. (2001). Drugs in the environment: emission of drugs, diagnostic aids and disinfectants into wastewater by hospitals in relation to other sources-a review. *Chemosphere.* 45 (6–7): 957–969.

28 Labadie, P., Cundy, A.B., Stone, K., et al. (2007). Evidence for the migration of steroidal estrogens through river bed sediments. *Environ Sci Technol.* 41 (12): 4299–4304.

29 Li, C.S. and Jenq, F.T. (1993). Physical and chemical composition of hospital waste. *Infect Control Hosp Epidemiol.* 14 (3): 145–150.

30 Luo, W., Lu, Y., Wang, G., et al. (2008). Distribution and availability of arsenic in soils from the industrialized urban area of Beijing, China. *Chemosphere.* 72 (5): 797–802.

31 Mahadevia, D. (2008). *Solid Waste Management in Indian Cities: Status and Emerging Practices.* Concept Publishing Company.

32 Mathur, P., Patan, S., and Shobhawat, A.S. (2012). Need of biomedical waste management system in hospitals-an emerging issue-a review. *Curr World Environ.* 7 (1): 117.

33 Mohankumar, S. and Kottaiveeran, K. (2007). Hospital waste management and environmental problems in India. *Framework.* 2 (6): 1621–1626.

34 Mubedi, J.I., Devarajan, N., Le Faucheur, S., et al. (2013). Effects of untreated hospital effluents on the accumulation of toxic metals in sediments of receiving system under tropical conditions: case of South India and Democratic Republic of Congo. *Chemosphere.* 93 (6): 1070–1076.

35 Muduli, K. and Barve, A. (2012). Challenges to waste management practices in Indian health care sector. In: *International Conference on Environment Science and Engineering. IPCBEE*, 32. Singapore: IACSIT Press.

36 Muthukannan, M. and Kumar, A. (2022). Influence of incinerated biomedical waste ash and waste glass powder on the mechanical and flexural properties of reinforced geopolymer concrete. *Aust J Struct Eng.* 23 (3): 254–268.

37 Nema, S.K. and Ganeshprasad, K.S. (2002). Plasma pyrolysis of medical waste. *Curr Sci.* 83 (3): 271–278.

38 Nematenda, P.P. (2020). *Compliance Level with Current Health Care Waste Management Regulation in Thulamela Local Municipality Health Care Establishments, South Africa.* South Africa: University of Venda.

39 Pan, M. and Chu, L.M. (2017). Fate of antibiotics in soil and their uptake by edible crops. *Sci Total Environ.* 599: 500–512.

40 Pan, M. and Chu, L.M. (2018). Occurrence of antibiotics and antibiotic resistance genes in soils from wastewater irrigation areas in the Pearl River Delta region, southern China. *Sci Total Environ.* 624: 145–152.

41 Pinto, V.N., Joshi, S.M., Velankar, D.H., et al. (2014). A comparative study of knowledge and attitudes regarding biomedical waste (BMW) management with a preliminary intervention in an academic hospital. *Int J Med Public Health.* 4 (1): 91–95.

42 Rezaee, A., Ansari, M., Khavanin, A., et al. (2005). Hospital wastewater treatment using an integrated anaerobic aerobic fixed film bioreactor. *Am J Environ Sci.* 1 (4): 259–263.

43 Shareefdeen, Z.M. (2012). Medical waste management and control. *J Environ Prot.* 3 (12): 1625.

44 Singh, H., Bhaskar, D.J., Dalai, D.R., et al. (2014). Dental biomedical waste management. *Int J Sci Study.* 2 (4): 66–68.

45 Singh, Z., Bhalwar, R., Jayaram, J., and Tilak, V.W. (2001). An introduction to essentials of bio-medical waste management. *Med J Armed Forces India.* 57 (2): 144–147.

46 Stringer, R., Kiama, J., Emmanuel, J., et al. (2010). *Non-Incineration Medical Waste Treatment Pilot Project at Bagamoyo District Hospital, Tanzania*, 37. Health Care Without Harm.

47 Sui, Q.m., Cao, X., Lu, S., et al. (2015). Occurrence sources and fate of pharmaceuticals and personal care products in the groundwater: a review. *Emerg Contam.* 1: 14–24.

48 Tiwari, A.V. and Kadu, P.A. (2013). Biomedical waste management practices in India-a review. *Int J Curr Eng Technol.* 3 (5): 2030–2033.

49 Van Boeckel, T.P., Brower, C., Gilbert, M., et al. (2015). Global trends in antimicrobial use in food animals. *Proc Natl Acad Sci USA.* 112 (18): 5649–5654.

50 Van Tran, T., Dai Cao, V., Nguyen, V.H., et al. (2020). MIL-53 (Fe) derived magnetic porous carbon as a robust adsorbent for the removal of phenolic compounds under the optimized conditions. *J Environ Chem Eng.* 8 (1): 102902.

51 Virmani, S., Nandigam, M., Kapoor, B., et al. (2017). Antibiotic use among health science students in an Indian university: a cross sectional study. *Clin Epidemiol Glob Health.* 5 (4): 176–179.

52 Willock, M. and Gabriel, D. (1998). Miscellaneous operational issues in the operating room. *Int Anesthesiol Clin.* 36 (1): 65–78.

53 Yadav, M. (2001). Hospital waste-a major problem. *JK Pract.* 8 (4): 276–282.

54 Zhou, L., Zhang, G., Tian, J., et al. (2017). Functionalized Fe_3O_4@ C nanospheres with adjustable structure for efficient hexavalent chromium removal. *ACS Sustain Chem Eng.* 5 (11): 11042–11050.

8

Valorization of Agricultural Wastes: Biological Processes and Techniques

8.1 Introduction

The continuous rise in human-induced greenhouse gas (GHG) emissions poses a significant threat to the stability of natural ecosystems and the sustainable advancement of human civilization [1]. Mitigating climate change by reducing GHG emissions remains a major challenge. Methane (CH_4) and nitrous oxide (N_2O) exhibit global warming potentials 23 and 298 times higher than carbon dioxide (CO_2), respectively [2], contributing approximately 17 and 6% to radiative forcing [3]. In China alone, the estimated annual emissions of CH_4 and N_2O from paddy soils range from 7.22 to 8.64 Tg and 88.01 to 98.11 Gg, respectively [4]. Given the environmental impact, a sustainable mitigation strategy for CH_4 and N_2O emissions in rice paddies is essential.

Agricultural residues, particularly crop straw, have been widely incorporated into farming systems to enhance soil fertility and organic carbon (C) stocks [5]. This practice helps curb pollution from straw burning [6] and plays a crucial role in regulating C and nitrogen (N) cycles [7]. However, direct application of agricultural waste in soil can exacerbate atmospheric GHG emissions [8]. For instance, straw incorporation in rice paddy fields has been shown to increase global warming potential significantly, with CH_4 emissions rising 3–11 times compared to control conditions [9]. Therefore, developing a strategy to minimize the GHG emissions associated with straw application is critical.

The valorization of agricultural waste through biochar production presents a promising alternative. Biochar, a stable C-rich material produced by the pyrolysis of plant residues under limited or zero oxygen, exhibits high pH, cation exchange capacity (CEC), and hydrophilic properties, along with significant porosity and surface area. These characteristics make biochar an effective soil amendment for improving soil fertility and nutrient dynamics [10]. Additionally, biochar has been investigated as a potential solution for regulating CH_4 and N_2O emissions in rice

Biological Processes and Techniques in Zero Waste Management, First Edition. Garima Awasthi and Mohit Nigam. Published 2026 by Wiley-VCH GmbH.

paddies [11]. For example, biochar addition to paddy soils under elevated temperature and CO_2 conditions reduced CH_4 emissions by 39.5%. However, conflicting findings suggest that while biochar application can reduce N_2O emissions, it may simultaneously enhance CH_4 emissions, likely due to increased microbial activity stimulated by additional C input [12]. Some studies have reported either neutral effects or increased biochar-induced GHG emissions, highlighting the dependency of biochar's impact on specific soil properties and conditions [12]. Despite its potential, the precise mechanisms underlying biochar-soil interactions concerning CH_4 and N_2O emissions remain insufficiently understood.

Over the past century, a 0.8°C rise in global surface temperatures has been primarily driven by the accumulation of long-lived GHGs. Increased human activities have significantly contributed to this warming, with CO_2, CH_4, and N_2O being the dominant forces behind global radiative forcing. These rising emissions pose severe environmental, ecological, and socio-economic challenges, potentially leading to substantial global mitigation costs. Since these three primary GHGs account for the majority of climate forcing, reducing their emissions is crucial for preventing further temperature escalation. Biochar, derived from agricultural waste, emerges as a valuable tool for climate change mitigation by enhancing C sequestration and improving soil health, making it a viable strategy for sustainable agricultural waste valorization.

8.1.1 Biochar Characteristics and Soil Impact

By applying biochar to soils, their physical, chemical, and biological characteristics may be significantly altered. These effects are entirely dependent on the biochar's characteristics, which are in turn influenced by its manufacturing technique. Biochar formed when biomass is heated to 400°C or higher has a minimum half-life of 100 years, whereas biochar produced at 600°C or above has a minimum half-life of 1,000 years. From these results, we conclude that the residence duration of biochar C in soil is crucial and that it may be used to sequester CO_2 from the atmosphere over long timescales.

The agricultural application of biochar has major benefits, including the reduction of soil GHG emissions and the absorption of nutrients or pollutants. Biochar possesses micrometer-scale holes that are big enough to hold nutrients, water, microorganisms, and other things. The surface area of biochar is larger than that of soil. Since biochar has a negative surface charge, it may attract positively charged molecules, such as ammonium ions (NH_4^+), and there is a potential benefit for the adsorption of inorganic N as fertilizer in agricultural soils.

In recent decades, several studies have explored in depth the effects of biochar application on soil characteristics [1]. Biochar application improves soil physical and chemical parameters, including pH, nutritional status, CEC, and soil stoichiometry [8]. According to a number of studies, biochar application enhances soil physical characteristics by increasing porosity, lowering bulk density, and promoting aggregation and water retention [13, 14]. Biochar's liming function decreases soil acidity by 31.9% (1.0–1.4 units for alkaline pine biochar) [1, 2]. Additionally, the use of biochar has favorable effects on CEC and electrical conductivity [3, 4]. Biochar may boost CEC by 20% and soil electrical conductivity by 124.6%, according to reports [2, 5]. The soil's water retention capacity is a crucial characteristic in relation to the soil's usable moisture content. Due to biochar's high porosity, researchers found improved soil water retention after biochar application. The majority of research revealed that biochar application enhanced soil water retention and reduced irrigation frequency [2]. In addition, biochar application modifies soil stoichiometry as a result of the change in biochar's organic C inputs. In certain temperate environments, increased soil organic C inputs encapsulate microbial N into biomass [7] and stabilize soil microbial N [3, 9]. The changes in soil dissolved organic C and NO_3 generated by biochar application are related to soil N_2O emissions [3]. Changes in soil physical structure and chemical characteristics resulting from biochar application are traceable to biochar properties, since biochar's porous structure enhances soil porosity and decreases bulk density [8]. These modifications would allow biochar to alter soil pH, improve water retention, and absorb and bind more nutrients in soil [13, 15].

8.2 Influence of Biochar on GHG Emissions

8.2.1 Soil N_2O Emission

There are several ways that N_2O may be created from soil. Nitrification, nitrifier denitrification, nitric acid (NO), ammonification, and denitrification are the microbial processes responsible for N_2O production in soils. Nitrification and denitrification are the primary sources of N_2O. In nitrification, ammonia (NH_3) or NH_4^+ is oxidized into nitrite (NO_2^-) or nitrate (NO_3^-), while in denitrification, NO_3^- is transformed into NO, N_2O, and dinitrogen dioxide (N_2O_2). The impact of biochar-amended soil on N_2O emissions may vary. Numerous short-term incubations may decrease N_2O emissions on biochar amendments; however, other short- and long-term experiments have not shown a reduction in soil N_2O emissions.

The following mechanism clarifies the impact of biochar amendment on soil N_2O emissions:

a. Biochar treatment improves soil aeration by reducing soil bulk density or by enhancing soil water retention. Aerated soil will affect nitrification and denitrification, reducing soil N_2O emissions.
b. Adsorption of soil inorganic N to the biochar surface immobilizes it, resulting in less N substrate for nitrifying and denitrifying enzymes and reduced N_2O emission.
c. The ratio of N_2O to N_2 emission decreases during denitrification at high soil pH.
d. The ratio of N_2O to N_2 denitrification products is decreased when substantial amounts of labile C are added to soil.
e. Inhibiting the action of soil nitrifying and denitrifying organisms due to volatile organic compounds and other chemicals produced by biochar.

8.2.2 Soil CO_2 Emissions

CO_2 release is limited by the soil's temperature, moisture, and substrate accessibility. The addition of biochar may affect soil CO_2 emissions [16].

The following processes elucidate the influence of biochar on soil CO_2 emissions:

a. The addition of biochar to soil increases soil aeration due to biochar's features, such as its smaller bulk thickness and better water-holding capacity. Releases of soil-derived CO_2 may be greatest in regions of "optimal soil moisture content" or increase when soil moisture content approaches saturation [12]. Biochar-induced soil disturbance enhances CO_2 emissions by remobilizing soil nutrients and microorganisms and increasing oxygen availability.

b. Biochar addition may increase soil microbial biomass due to the expansion of the C-use proficiency of the framework after aggregation of soil organic carbon (SOC), organisms, and supplements onto the surface of the biochar. It is possible that the mobility of C-mineralizing proteins might be reduced during biochar transformation, reducing soil CO_2 emissions. Finally, it has been postulated that soil-derived CO_2 may adsorb to the surface of biochar as carbonates, reducing its outflow to the atmosphere [3].

8.2.3 CH_4 Emissions from Soil

Studies conducted in rice fields demonstrate that biochar modification affects soil CH_4 emissions. Rice paddies emit far more CH_4 than aerobic agricultural soils. Wang et al. discovered that altering rice paddies with biochar enhanced soil CH_4 emissions by 31%. The hypothesized methods for clarifying the influence of biochar on soil CH_4 emissions are as follows:

a. Labile C is added to biochar-amended soil, where it serves as a substrate for methanogens and increases soil CH_4 generation.

b. Biochar-amended soil is strongly aerated, functions as a natural sink for atmospheric CH_4, and oxidizes CH_4 to CO_2 as a by-product.

8.3 Effects of Biochar Addition on the Composition and Activity of Soil Microorganisms

Biochar application increases soil physical, chemical, and microbiological characteristics [8]. Biochar's vast surface area and reactivity attract ions and organic molecules, increasing the number of possible sites for microbe–substrate interactions [1]. Therefore, biochar application has the ability to alter soil microbial community composition, microbial activity, and soil enzyme activity, therefore impacting several biogeochemical processes [2]. Biochar treatment often alters soil microbial communities toward beneficial fungi and rhizobacteria that promote plant development [16]. It was discovered that biochar treatment enhances the number of phosphorus (P) mobilizing bacterial species [2]. The use of biochar also increases the number of live bacteria cells within root nodules [3]. It was discovered that biochar application dramatically increases the ratio of Gram-positive to Gram-negative bacteria and results in high CO_2 emissions [4]. In addition, biochar treatment reduces the quantity of mycorrhizal fungus by lowering the nutritional and water needs of the mycorrhizal symbiosis and has direct detrimental impacts on mycorrhizal development [2, 3]. It was discovered that biochar enhances the colonization of roots by arbuscular mycorrhizal (AM) fungi [3]. It was also discovered favorable effects on AM fungal root colonization since biochar promotes the germination of AM spores. Application of biochar promotes zoospore-forming pathogens [4]. Biochar may stimulate soil microorganisms and boost their activity [3]. It was observed that biochar application boosts the respiration activity of soil microbes. Similarly, it was found that biochar treatment increases basal respiration by 30.1%. However, the addition of biochar has the opposite impact on microbial activity [17]. Biochar inhibits dehydrogenase and esterase activity by absorbing or inhibiting substrates [18].

Changes in soil microbial community composition, biomass, and activity might potentially be linked to biochar's physical and chemical characteristics as well as biochar-induced changes in soil parameters [19]. Biochar, for instance, may offer easily accessible nutrients for soil microorganisms, and the addition of a minor labile component can boost microbial activity [12, 16].

8.4 The Impact of Biochar on Soil Nutrients

Biochar addition modifies nutrient cycles by directly contributing nutrients (N, P, K, Ca, and Mg) in biochar to the soil [3] and by indirectly altering nutrient exchange owing to an increased reactive surface area [13]. The effects of biochar treatment on N dynamics and N-cycling enzymes remain unclear [17], since soil N mineralization was observed to decrease [7], increase [8], and stay constant [5] after biochar application. Biochar addition primarily changes N fluxes in soil and lowers N emissions as a gas [5]. Biochar boosts nitrification activity in forest soils but not in grassland soils and increases CEC, according to studies [6]. Biochar modifies the dynamics of soil nitrogenase and accelerates N dynamics [6]. It has been shown that biochar treatment affects the development of nodules and N fixation. Biochar of diverse sources has varying effects on soil nutrients since the nutritional content of biomass raw materials varies [3]. For instance, biochar derived from *Miscanthus* may be used as a slow-release silicon fertilizer [4], but biochar derived from animal bones is an effective P fertilizer [6]. Due to the adsorption of orthophosphate and organic P molecules on its surface, biochar may reduce the bioavailability of P. In contrast, biochar reduces nutrient leaching [6]. As a consequence, biochar-enriched soils contain more accessible P for plant growth [9].

8.5 Influence of Biochar Addition on Soil GHG Emissions and C Sequestration

Biochar application may ameliorate climate change by lowering GHG emissions and improving soil C storage [10], although the precise effects of biochar application on soil GHG emissions are still debated and vary according to research [3]. In some studies [17], soil CO_2, CH_4, and N_2O fluxes rise greatly, but in others [8], they drop dramatically or stay stable. Biochar application to paddy soils, for instance, results in a 13% rise in CO_2 emissions but a 41.8% reduction in N_2O emissions [11]. Another field experiment revealed that biochar treatment had no substantial influence on soil CO_2 and N_2O emissions in a grazing environment [2]. As a result of these varying effects on soil GHG emissions, the consequences of biochar application are still uncertain. Increases in soil CO_2 fluxes have been reported after biochar application. Biochar application may help to mitigate climate change by lowering GHG emissions and improving soil C storage [1], though the exact effects of biochar application on soil GHG emissions are still being debated and vary between studies [12, 16]. In some studies [17], soil CO_2, CH_4, and N_2O fluxes rise greatly, whereas in others [7] they decrease or stay stable. Biochar application to paddy soils, for instance, results in a 13% rise in CO_2 emissions but a 41.8% reduction in N_2O emissions [11]. Another field experiment revealed that biochar treatment had no substantial influence on soil CO_2 and N_2O emissions in a grazing environment [12]. Due to these varying effects on soil GHG emissions, the consequences of biochar application are still uncertain. Several investigations have indicated increases in soil CO_2 fluxes due to biochar application, although with substantial uncertainty [16]. In some research (e.g., cited in a number of studies, but with substantial uncertainty), biochar amendment reduced soil CO_2 fluxes or had no impact [16]. In other investigations (e.g., [15, 18]), the addition of biochar reduced soil CO_2 fluxes or had no impact. Changes in soil CO_2 emission rates may be influenced by a number of variables, including soil temperature, soil type, disturbance intensity, biochar pyrolysis temperature, and biochar application rates [14, 19]. This increase in soil CO_2 emissions following application of biochar may be the consequence of a variety of processes, including (i) biotic use of biochar and breakdown of biochar [20], (ii) release of biochar-C such as carbonates [12], and (iii) priming of native soil organic C pools owing to interactions between biochar and native soil organic matter [11]. In addition, it has been shown that biochar addition considerably enhances the ratios of Gram-positive and Gram-negative bacteria and may contribute to higher CO_2 fluxes from organic C breakdown in soil. Changes may also result from the direct addition of highly unstable C to biochar [16].

8.6 Biochar Production

In South America's Amazon basin, charcoal-rich soils were found in the twentieth century [16]. These "Amazonian Dark Earths" were the result of hundreds of years of human management and included much higher levels of charcoal-derived C, SOC, and nutrients than neighboring soils [11]. Biochar refers to charcoal manufactured in a controlled setting and applied to soil [16].

Initial studies indicate that biochar may function as an agent for C sequestration. The addition of charcoal generated in a controlled environment to soil is referred to as "biochar." Pyrolysis is the process of producing it by heating it to temperatures ranging from 350 to 600°C in an anaerobic or oxygen-limited environment. Similar

Table 8.1 Various biochar manufacturing methods.

Method	Heating input (°C)	Time	Biochar yield (%)	Gas products (CH_4, CO, H_2) (%)
Slow pyrolysis	350–400	2–30 min	35	35
Intermediate pyrolysis	350–450	1–15 min	20	25
Fast pyrolysis	450–550	1–5 s	12	13
Gasification	>750	10–20 s (vapors)	10	About 80

to charcoal, it has a high C concentration, a low N content, a large surface area, and cation exchange ability. The processes of pyrolysis, hydrothermal carbonization, and microwave carbonization are used for the preparation and manufacturing of biochar. The preparation of biochar is shown in Table 8.1. The four types of pyrolysis processes used to process raw materials are as follows: (i) slow; (ii) rapid; (iii) flash; and (iv) gasification.

Four basic heating temperatures, times of heating, and biochar yields characterize the four essential biochar manufacturing operations. The procedure is summarized in Table 8.1.

8.7 Conclusion

The valorization of agricultural waste through biochar production offers a sustainable approach to enhancing soil quality, improving crop productivity, and mitigating climate change. Biochar application has demonstrated significant benefits for soil physical, chemical, and biological properties, influencing bulk density, pH, water retention capacity, and nutrient availability. Although variations in biochar characteristics, application methods, and dosage rates lead to diverse outcomes across studies, the majority of research highlights its positive impacts on soil fertility, crop yields, and GHG emissions reduction. The beneficial effects of biochar can be attributed to several key mechanisms, including improved nutrient retention and soil water-holding capacity, enhanced microbial activity, reduced soil CO_2 emissions through C stabilization, and decreased N_2O emissions via improved soil aeration and pH balance. By integrating biochar into agricultural systems, not only can soil health be restored, but carbon sequestration can also be enhanced, contributing to climate resilience. Overall, the transformation of agricultural residues into biochar aligns with the principles of circular economy and environmental sustainability. As research continues to refine biochar application techniques, its role in sustainable agriculture will become increasingly vital, making it a key strategy for optimizing resource utilization and reducing the environmental footprint of agroecosystems.

References

1 Smith, P. and Fang, C.M. (2010). A warm response by soils. *Nature*. 464: 499–500. https://doi.org/10.1038/464499a.

2 Munoz, C., Paulino, L., Monreal, C., and Zagal, E. (2010). Greenhouse gas (CO_2 and N_2O) emissions from soils: a review. *Chilean J Agric Res*. 70: 485–497. https://doi.org/10.4067/S0718-58392010000300016.

3 WMO (2017). *World Meteorological Organization Greenhouse Gas Bulletin: The State of Greenhouse Gases in the Atmosphere Based on Observations Through 2016*. Geneva: WMO.

4 Xing, G.X. (1998). N_2O emission from cropland in China. *Nutr Cycle Agroecosyst*. 52: 249–254. https://doi.org/10.1023/a:1009776008840.

5 Li, C.S., Mosier, A., Wassmann, R., et al. (2004). Modeling greenhouse gas emissions from rice-based production systems: sensitivity and upscaling. *Glob Biogeochem Cycles*. 18: GB1042. https://doi.org/10.1029/2003gb002045.

6 Kharub, A.S., Sharma, R.K., Mongia, A.D., et al. (2004). Effect of rice (*Oryza sativa*) straw removal, burning and incorporation on soil properties and crop productivity under rice-wheat (*Triticum aestivum*) system. *Indian J Agric Sci*. 74: 295–299.

7 Lugato, E., Berti, A., and Giardini, L. (2006). Soil organic carbon (SOC) dynamics with and without residue incorporation in relation to different nitrogen fertilisation rates. *Geoderma*. 135: 315–321. https://doi.org/10.1016/j.geoderma.2006.01.012.

8 Zou, J.W., Huang, Y., Lu, Y.Y., et al. (2005). Direct emission factor for N_2O from rice-winter wheat rotation systems in southeast China. *Atmos Environ*. 39: 4755–4765. https://doi.org/10.1016/j.atmosenv.2005.04.028.

9 Ma, J., Li, X.L., Xu, H., et al. (2007). Effects of nitrogen fertiliser and wheat straw application on CH_4 and N_2O emissions from a paddy rice field. *Aust J Soil Res*. 45: 359–367. https://doi.org/10.1071/SR07039.

10 Bakar, R.A., Razak, Z.A., Ahmad, S.H., et al. (2015). Influence of oil palm empty fruit bunch biochar on floodwater pH and yield components of rice cultivated on acid sulphate soil under rice intensification practices. *Plant Prod Sci*. 18: 491–500. https://doi.org/10.1626/pps.18.491.

11 He, Y.H., Zhou, X.H., Jiang, L.L., et al. (2017). Effects of biochar application on soil greenhouse gas fluxes: a meta-analysis. *Glob Change Biol Bioenergy*. 9: 743–755. https://doi.org/10.1111/gcbb.12376.

12 Singla, A. and Inubushi, K. (2014). Effect of biochar on CH_4 and N_2O emission from soils vegetated with paddy. *Paddy Water Environ*. 12: 239–243. https://doi.org/10.1007/s10333-013-0357-3.

13 Ding, Z.H., Wan, Y.S., Hu, X., et al. (2016). Sorption of lead and methylene blue onto hickory biochars from different pyrolysis temperatures: importance of physicochemical properties. *J Ind Eng Chem*. 37: 261–267. https://doi.org/10.1016/j.jiec.2016.03.035.

14 Baiamonte, G., De Pasquale, C., Marsala, V., et al. (2015). Structure alteration of a sandy-clay soil by biochar amendments. *J Soils Sediment.* 15: 816–824. https://doi.org/10.1007/s11368-014-0960-y.

15 Arif, M., Ilyas, M., Riaz, M., et al. (2017). Biochar improves phosphorus use efficiency of organic-inorganic fertilizers, maize-wheat productivity and soil quality in a low fertility alkaline soil. *Field Crops Res.* 214: 25–37. https://doi.org/10.1016/j.fcr.2017.08.018.

16 Lehmann, J. (2007). Bio-energy in the black. *Front Ecol Environ.* 5: 381–387. https://doi.org/10.1890/1540-9295(2007)5[381:BITB]2.0.CO;2.

17 Xiang, Y.Z., Deng, Q., Duan, H.L., and Guo, Y. (2017). Effects of biochar application on root traits: a meta-analysis. *Glob Change Biol Bioenergy.* 9: 1563–1572. https://doi.org/10.1111/gcbb.12449.

18 Lehmann, J., Rillig, M.C., Thies, J., et al. (2011). Biochar effects on soil biota—a review. *Soil Biol Biochem.* 43: 1812–1836. https://doi.org/10.1016/j.soilbio.2011.04.022.

19 Van Zwieten, L., Kimber, S., Morris, S., et al. (2010). Effects of biochar from slow pyrolysis of papermill waste on agronomic performance and soil fertility. *Plant Soil.* 327: 235–246. https://doi.org/10.1007/s11104-009-0050-x.

20 Oguntunde, P.G., Fosu, M., Ajayi, A.E., and van de Giesen, N. (2004). Effects of charcoal production on maize yield, chemical properties and texture of soil. *Biol Fertil Soils.* 39: 295–299. https://doi.org/10.1007/s00374-003-0707-1.

9

Biological Methods and Techniques for Plastic Waste Valorization

9.1 Introduction

Plastics have become an indispensable part of modern society due to their versatility, durability, and cost-effectiveness. From packaging materials and household items to medical equipment and automotive components, plastics are embedded in nearly every sector of contemporary life. However, this convenience has come at a significant environmental cost. The proliferation of plastic production, coupled with inadequate waste management systems, has led to an unprecedented global plastic waste crisis. As concerns regarding environmental sustainability and public health continue to mount, there is an urgent need to explore alternative, eco-friendly strategies for managing and repurposing plastic waste. Among the emerging solutions, biological valorization has gained considerable attention as a sustainable and innovative pathway for converting plastic waste into valuable resources. Since the onset of mass production in the 1950s, plastic manufacturing has grown exponentially. As of 2024, global plastic production exceeds **400 million tonnes annually**, with projections indicating continued growth due to expanding demands in packaging, construction, consumer goods, and healthcare. Alarmingly, a significant proportion, estimated at **over 40%**, comprises single-use plastics designed for short-term utility but long-term persistence in the environment [1].

Despite the scale of production, plastic waste management remains inefficient and inconsistent across regions. Recent global estimates suggest that only 9–10% of plastic waste is effectively recycled, around 12% is incinerated for energy recovery, and the remaining nearly 80% is either landfilled or leaked into natural environments. Developing nations face particular challenges due to inadequate infrastructure, informal waste sectors, and limited public awareness. Even in industrialized countries, recycling systems often fail to process mixed, contaminated, or low-grade plastics efficiently, leading to stockpiling or exportation to countries with weaker environmental regulations. One of the most pressing issues with plastic waste is its exceptional persistence. Conventional plastics, primarily derived from fossil

Biological Processes and Techniques in Zero Waste Management, First Edition. Garima Awasthi and Mohit Nigam. Published 2026 by Wiley-VCH GmbH.

fuels, such as polyethylene (PE), polypropylene (PP), polystyrene (PS), and polyethylene terephthalate (PET), are engineered for durability. These materials do not readily degrade under natural conditions and can persist in the environment for hundreds of years. Instead of decomposing, they fragment into smaller particles, microplastics (less than 5 mm) and nanoplastics, which spread widely across terrestrial, freshwater, and marine ecosystems. The ecological consequences of plastic pollution are profound. Microplastics have been detected in soil, rivers, oceans, atmospheric air, and even human blood and placentas. Wildlife, including birds, fish, and mammals, can ingest or become entangled in plastic debris, leading to injury, suffocation, or death. Moreover, plastics often contain or adsorb hazardous substances like bisphenol A (BPA), phthalates, heavy metals, and persistent organic pollutants (POPs), which can disrupt hormonal systems and pose serious health risks. These pollutants can also bioaccumulate along the food chain, ultimately impacting human populations [2].

To mitigate the environmental impacts of plastic waste, traditional waste management strategies have been implemented globally, including landfilling, incineration, and mechanical and chemical recycling. However, each of these approaches carries inherent limitations that hinder their long-term sustainability and effectiveness.

- **Landfilling:** The most commonly practiced disposal method is often the default option in regions with limited waste processing capabilities. Although convenient and inexpensive in the short term, landfilling leads to long-term land use issues and environmental degradation. Plastics in landfills do not biodegrade and may leach harmful additives into the surrounding soil and groundwater.
- **Incineration:** Reduces the volume of plastic waste and can generate energy, but it is a controversial method due to the emission of greenhouse gases and toxic air pollutants such as dioxins and furans. While modern incinerators equipped with advanced filtration systems mitigate some risks, the high cost of installation and operation, coupled with public opposition, limits widespread adoption.
- **Mechanical Recycling:** Involves sorting, cleaning, shredding, and remelting plastics into new products. While relatively straightforward, it suffers from significant drawbacks, including material degradation after repeated cycles, challenges in separating multilayered or contaminated plastics, and low economic returns for recycled plastics.
- **Chemical Recycling:** Technologies attempt to break down plastics into monomers or fuels using thermal or catalytic processes. Though promising in theory, these technologies remain costly, energy-intensive, and technologically immature. Moreover, their environmental benefits over conventional recycling are still debated, particularly when fossil fuel-derived energy is used [3].

These traditional methods, while necessary in the short term, are insufficient to address the scale and complexity of global plastic waste. As such, a paradigm shift toward more sustainable and circular solutions is urgently needed.

- **Biological Valorization:** Represents a transformative approach to plastic waste management by leveraging the power of living systems, such as microorganisms and enzymes, to degrade, transform, or repurpose plastics into valuable products. Unlike conventional mechanical or chemical recycling, biological methods aim to operate under mild conditions, consume less energy, and generate fewer secondary pollutants, making them inherently more sustainable and environmentally friendly [4].

This involves multiple strategies, including:

- **Biodegradation:** Where microbes or enzymes break down plastics into simpler compounds.
- **Bioconversion:** Where those breakdown products are transformed into bio-based chemicals or fuels.
- **Bio-upcycling:** Where degraded plastics are used as feedstock to produce higher-value bioproducts like biosurfactants, bioplastics, or pharmaceuticals.

Recent breakthroughs in **synthetic biology** and **genetic engineering** have opened up exciting possibilities for enhancing the plastic-degrading capabilities of microbes. Notable examples include *Ideonella sakaiensis*, a bacterium that produces PETase and MHETase enzymes capable of depolymerizing PET into its constituent monomers. Other organisms, such as species from the genera *Pseudomonas, Bacillus*, and *Aspergillus*, have shown potential in degrading various synthetic polymers and converting them into biochemicals or biomass. Moreover, **metagenomic approaches** and **microbial consortia engineering** are now being explored to identify and optimize communities of microorganisms that can act synergistically to degrade mixed plastic waste streams. These biological systems can be integrated into **waste biorefineries**, enabling the co-processing of plastic and organic wastes into a suite of renewable products. Despite the promise, several challenges remain, such as low degradation efficiency, incomplete mineralization, scalability, and biosafety concerns. Addressing these hurdles will require interdisciplinary research, public–private collaboration, and supportive regulatory frameworks [5].

9.2 The Need for Biological Valorization of Plastics

The increasing accumulation of plastic waste in the environment has become one of the most pressing global challenges of the twenty-first century. Conventional methods of plastic waste management such as landfilling, incineration, and mechanical recycling have proven to be insufficient, unsustainable, and often environmentally harmful. Landfilling leads to long-term soil and groundwater pollution, while incineration emits greenhouse gases and toxic pollutants. Mechanical recycling, although widely practiced, is limited by material contamination, polymer degradation, and low economic viability. As a result, vast quantities of plastic continue to persist in natural ecosystems, causing severe ecological and health hazards.

In this context, biological valorization offers a promising alternative that aligns with the principles of sustainability and the circular economy. This approach harnesses the capabilities of microorganisms, enzymes, or other biological systems to break down and convert plastic waste into useful, high-value products such as biofuels, bioplastics, organic acids, or other industrially relevant compounds. Unlike traditional methods, biological processes often operate under mild environmental conditions, produce fewer secondary pollutants, and can potentially handle mixed or low-grade plastic waste that is otherwise unrecyclable. Moreover, advances in biotechnology, synthetic biology, and systems biology have opened new avenues for engineering microbial strains and enzymatic pathways to enhance the efficiency of plastic degradation and conversion. By turning a persistent pollutant into a valuable resource, biological valorization not only helps in mitigating plastic pollution but also contributes to resource recovery, climate change mitigation, and sustainable industrial development. Thus, there is a compelling need to invest in and scale up biological strategies for plastic waste valorization to support global environmental and economic sustainability goals [6].

9.3 Types of Plastics and Their Biodegradability

Plastics are a ubiquitous part of modern life, valued for their durability, flexibility, and cost-effectiveness. However, the same characteristics that make plastics so useful also contribute to their persistence in the environment. Understanding the types of plastics, both conventional and biodegradable, is critical for evaluating their environmental impacts and exploring strategies for sustainable management and valorization. This section explores the major types of plastics, their molecular characteristics, and the varying degrees of biodegradability they exhibit, including a focus on microplastics and nanoplastics and the challenges they pose [7].

- **Conventional Plastics: PE, PET, PS, PVC, PP**
  Conventional or petroleum-based plastics are synthesized from nonrenewable fossil fuels and are designed to resist degradation. Among the most commonly used conventional plastics are PE, PET, PS, polyvinyl chloride (PVC), and PP.

  PE, available in high-density (HDPE) and low-density (LDPE) forms, is extensively used in packaging, bags, containers, and films. Its chemical structure comprises long chains of ethylene monomers with minimal branching in HDPE and more branching in LDPE. This structure renders PE highly resistant to microbial attack due to its hydrophobic nature, high molecular weight, and lack of functional groups that can be enzymatically cleaved.

  PET is widely used in beverage bottles and food containers. Its semicrystalline structure and aromatic backbone make it resistant to biodegradation. Although PET can be degraded to some extent by certain enzymes like PETases and MHETases, natural degradation rates remain extremely slow, particularly in environmental conditions that lack these enzymes.

PS, commonly found in disposable cutlery, insulation, and packaging materials, is another persistent plastic. Its aromatic phenyl groups provide rigidity but also inhibit microbial degradation. Even in composting or landfill environments, PS degradation is negligible over decades.

Polyvinyl chloride (PVC) is used in pipes, flooring, and medical devices. Its backbone contains chlorine atoms, which not only increase its mechanical strength and fire resistance but also make it toxic and resistant to microbial degradation. Moreover, PVC often contains a mix of plasticizers, stabilizers, and heavy metals that further complicate its biodegradability and environmental safety.

PP, used in textiles, automotive parts, and packaging, has a high melting point and excellent resistance to chemical and microbial attack. Its high hydrophobicity and crystalline structure hinder biodegradation, although some bacteria and fungi have shown limited ability to degrade it under specialized conditions.

Overall, conventional plastics are designed for longevity, which makes their widespread disposal particularly problematic. Most of them degrade very slowly, leading to accumulation in landfills and aquatic environments, where they fragment into microplastics but do not truly biodegrade.

- **Bioplastics and Biodegradable Polymers: PLA, PHB, PCL, PBS**

  In response to the environmental challenges posed by conventional plastics, bioplastics and biodegradable polymers have gained significant attention. These materials are either derived from renewable sources or designed to degrade more readily in the environment. Some of the notable examples include polylactic acid (PLA), polyhydroxybutyrate (PHB), polycaprolactone (PCL), and polybutylene succinate (PBS) [8].

  PLA is a thermoplastic aliphatic polyester derived from renewable resources like cornstarch or sugarcane. It is biodegradable under industrial composting conditions, where high temperatures and microbial activity can break down its ester bonds. PLA finds extensive use in packaging, 3D printing, and disposable items. However, its biodegradation is limited under ambient or marine conditions, requiring specific environmental parameters for complete breakdown.

  PHB is a type of polyhydroxyalkanoate (PHA) produced by microbial fermentation of sugars or lipids. It is completely biodegradable and biocompatible, decomposing into CO_2 and water under both aerobic and anaerobic conditions. PHB exhibits thermoplastic properties similar to PP and is used in medical applications and packaging. However, its high production cost and brittleness limit its widespread adoption.

  PCL is a synthetic biodegradable polyester with a low melting point and excellent blend compatibility with other polymers. It degrades relatively slowly under environmental conditions but is completely biodegradable in composting settings. Its degradation occurs via hydrolysis of ester linkages, which are then further metabolized by microorganisms. PCL is used in controlled drug delivery, tissue engineering, and biodegradable packaging.

PBS is another aliphatic polyester with good thermal and mechanical properties. Derived from succinic acid and 1,4-butanediol, PBS is biodegradable in soil and compost environments. It is gaining popularity in applications like agricultural films, food packaging, and single-use items. PBS biodegrades via enzymatic action and is considered one of the promising candidates for replacing traditional fossil-based plastics.

Bioplastics like PLA, PHB, PCL, and PBS demonstrate varying degrees of biodegradability depending on their molecular structure and environmental conditions. While these materials are not a silver bullet, they represent a significant step toward sustainable plastic alternatives, especially when integrated with circular economy practices.

- **Crystalline vs. Amorphous Structures and Their Degradation Behavior**
  The molecular arrangement of polymers, whether crystalline or amorphous, plays a significant role in determining their biodegradability. Crystalline regions are tightly packed, highly ordered structures that are less accessible to water and enzymes, making them more resistant to degradation. In contrast, amorphous regions are disordered and loosely packed, allowing for easier penetration of water molecules and enzymatic action. For example, in semicrystalline polymers like PET and PLA, degradation typically begins in the amorphous regions where enzymes can access the polymer chains. As degradation progresses, the crystalline zones remain intact for longer periods, significantly slowing down the overall process. Similarly, highly crystalline plastics like HDPE and PP show very low biodegradability due to their structural inaccessibility. Manipulating the degree of crystallinity through polymer synthesis or post-processing techniques is a strategy being explored to enhance biodegradability without compromising material properties. However, this approach must be carefully balanced to retain the functional utility of the polymer while improving its environmental performance [9].
- **Microplastics and Nanoplastics: Challenges in Degradation**
  As plastics degrade physically through mechanical abrasion, UV exposure, or thermal stress, they break down into smaller fragments known as microplastics (less than 5 mm) and nanoplastics (less than 100 nm). These fragments do not biodegrade readily and instead persist in the environment, entering food chains and posing significant ecological and health risks. Microplastics originate from both primary sources (such as cosmetic microbeads and industrial abrasives) and secondary sources (fragmentation of larger plastic debris). Nanoplastics are even more concerning due to their size, which allows them to cross biological membranes and accumulate in organs and tissues.

  The degradation of micro- and nanoplastics is particularly challenging because of their small size, high surface area-to-volume ratio, and the presence of additives like plasticizers, flame retardants, and stabilizers. These materials often adsorb POPs, heavy metals, and other toxicants, compounding their environmental toxicity. Moreover, their small size makes collection and removal from natural environments exceedingly difficult. While certain microbes have shown promise in degrading specific types of microplastics, the efficiency of such processes remains limited and is an active area of research. Advanced strategies involving enzyme engineering, microbial consortia, and hybrid oxidation–biological systems are being explored to address these challenges [10].

9.4 Microbial Systems in Plastic Degradation

Microbial systems play a crucial role in the biodegradation of plastics by transforming persistent synthetic polymers into simpler, environmentally benign molecules. These systems rely on the metabolic capabilities of bacteria, fungi, or microbial consortia to break down complex plastic structures into monomers or mineralized forms such as CO_2 and H_2O. Understanding the microbial players involved, the mechanisms they utilize, and how they function individually or in communities is essential to designing effective biotechnological solutions for plastic waste valorization. This section focuses on three major components: bacterial systems, fungal systems, and consortia-based approaches, each of which contributes uniquely to the plastic degradation process [11].

9.4.1 Bacterial Systems

Bacterial systems are among the most studied and versatile tools in plastic biodegradation. Species like *Pseudomonas* spp., *I. sakaiensis*, and *Bacillus* spp. have shown strong potential for degrading various types of plastics, including PE, PET, PS, and polyurethane (PU). These bacteria utilize specific enzymatic mechanisms that allow them to adhere to plastic surfaces, initiate depolymerization, and mineralize the breakdown products. *Pseudomonas* spp. are particularly well known for their ability to degrade polyolefins and PUs. They produce extracellular enzymes such as oxygenases and esterases that cleave the polymer chains into oligomers and monomers. For instance, *Pseudomonas putida* can metabolize monomers like terephthalic acid (TPA), derived from PET degradation, and convert them into PHAs, a class of biodegradable plastics. Similarly, *Bacillus subtilis* and *Bacillus cereus* have demonstrated capabilities to degrade LDPE through the production of lipases and proteases.

A remarkable breakthrough in bacterial plastic degradation came with the discovery of *I. sakaiensis*, a bacterium isolated from a PET-contaminated wastewater site in Japan. This bacterium secretes two key enzymes: PETase and MHETase. PETase hydrolyzes PET into mono-(2-hydroxyethyl) terephthalate (MHET), which is further broken down by MHETase into TPA and ethylene glycol (EG). These monomers are then taken up by the bacterial cells and used as carbon and energy sources. The degradation process involves an initial phase of plastic surface colonization, followed by the secretion of depolymerizing enzymes, uptake of intermediates, and their assimilation through central metabolic pathways [12].

This overall process, adhesion, depolymerization, and mineralization, is a hallmark of bacterial plastic degradation. Bacteria first form biofilms or surface attachments that help them overcome the hydrophobic barrier of plastics. Then, secreted enzymes initiate chemical attacks on polymer chains, creating fragments small enough for cellular uptake. Inside the cells, these fragments enter metabolic routes like the TCA cycle or β-oxidation for complete breakdown.

9.4.2 Fungal Systems

Fungi also play a pivotal role in the biodegradation of synthetic polymers, particularly through their extensive enzyme arsenal and hyphal growth pattern that aids in plastic surface colonization. Species such as *Aspergillus* spp., *Penicillium* spp., and *Fusarium* spp. are widely studied for their ability to degrade plastics like PE, PVC, and PU. *Aspergillus niger* and *Aspergillus tubingensis*, for example, produce oxidative and hydrolytic enzymes, including laccases, cutinases, and esterases, which are effective in cleaving ester bonds in synthetic polymers. These enzymes reduce the molecular weight of plastic substrates, increasing their susceptibility to further degradation. *Penicillium chrysogenum* and *Fusarium solani* exhibit similar capabilities, with some studies reporting weight loss in PE films exposed to fungal cultures for several weeks. A unique feature of fungal degradation is **hyphal penetration**, where fungal hyphae physically invade the plastic matrix. This not only increases the surface area of enzymatic contact but also facilitates mechanical disruption of the polymer, allowing deeper enzymatic action. Fungi thrive in varied environments and can tolerate harsh conditions, making them suitable candidates for bioaugmentation in contaminated soils or waste dumps [13].

The enzymatic degradation process typically begins with **biodeterioration**, where fungal enzymes weaken the plastic structure by oxidizing surface bonds. This is followed by **bio-fragmentation**, in which large polymer chains are broken down into oligomers and monomers. Finally, **assimilation** occurs as the fungus absorbs these monomers, which are metabolized via central pathways for energy and biomass synthesis. These processes can be optimized by adjusting environmental parameters such as pH, moisture, and nutrient availability.

9.4.3 Consortia and Community-based Approaches

While individual microbial strains can degrade plastics to varying extents, microbial consortia, the mixed communities of bacteria and/or fungi, have demonstrated significantly enhanced degradation performance due to their synergistic interactions. In natural and engineered systems, microbial consortia enable **division of labor**, where different members specialize in distinct tasks: one group may degrade the plastic into oligomers, another may mineralize these fragments, and yet another may produce biosurfactants that increase plastic accessibility. Biofilms formed by microbial consortia play a crucial role in this process. Biofilms are structured microbial communities embedded in a self-produced extracellular polymeric substance (EPS) matrix that adheres strongly to plastic surfaces. The biofilm environment facilitates close cell-to-cell communication, gene exchange, and efficient substrate channeling. For example, in a mixed culture of *Pseudomonas, Bacillus*, and *Rhodococcus* species, each bacterium may secrete different enzymes targeting various bonds in a plastic polymer, resulting in faster and more complete degradation compared to monocultures. One study demonstrated that a bacterial consortium isolated from plastic-contaminated soil degraded LDPE films more efficiently than individual strains, achieving a weight reduction of over 15% in 60 days. Another example involved a co-culture of *Aspergillus* and *Pseudomonas*, where fungal hyphae enhanced surface degradation while bacterial enzymes metabolized the released fragments. These results suggest that consortia can tackle complex, heterogeneous plastic waste streams more effectively [14].

The degradation process in consortia generally follows a multistep route: (i) Initial biofilm formation and surface attachment; (ii) Collaborative enzymatic depolymerization of the plastic matrix; (iii) Metabolic handover, where intermediate compounds are shared among community members; and (iv) Complete mineralization or conversion into secondary products such as CO_2, biomass, or bioplastics. These systems mimic natural degradation pathways and are increasingly being considered for application in bioreactors and integrated waste treatment systems.

9.5 Enzymatic Processes for Plastic Valorization

The persistent and nonbiodegradable nature of synthetic plastics has emerged as one of the most pressing environmental concerns of the twenty-first century. While physical and chemical recycling methods have been implemented at various scales, they often result in downcycling, are energy-intensive, and produce secondary pollutants. In contrast, enzymatic plastic degradation and valorization offer a green, specific, and sustainable alternative. Enzymes, as biological catalysts, have the capacity to selectively break down polymer chains into monomers or oligomers that can be reutilized for the synthesis of value-added products. This process not only helps reduce plastic pollution but also supports circular economy principles. This section delves into key enzymes involved in plastic degradation, their mechanisms of action, kinetic properties, technological innovations like immobilization systems, and strategies for enhancing enzymatic activity through engineering and directed evolution.

Key enzymes in plastic degradation: PETase, MHETase, cutinase, laccase, and lipase. Several enzymes have shown promise in the biodegradation of synthetic polymers, particularly PET and polyesters. Among them, **PETase**, **MHETase**, **cutinase**, **laccase**, and **lipase** are among the most studied [15].

- **PETase (Polyethylene Terephthalate Hydrolase)** Discovered in *I. sakaiensis*, PETase is a specialized enzyme capable of hydrolyzing PET into its primary monomers—TPA and MHET. PETase functions by attacking the ester bonds in the polymer backbone, particularly at amorphous regions of PET, which are more accessible than crystalline zones. PETase's active site resembles those of cutinases, with a broader catalytic cleft that accommodates PET chains more effectively. Its discovery marked a major milestone in plastic biodegradation, and subsequent studies have sought to improve its activity and thermal stability.

- **MHETase (Mono(2-Hydroxyethyl) Terephthalate Hydrolase)** It, also from *I. sakaiensis*, acts synergistically with PETase by converting MHET, one of the intermediate products of PET degradation, into TPA and EG, both of which are commercially valuable. MHETase ensures complete breakdown of PET-derived intermediates, thereby increasing the efficiency of enzymatic recycling. Its crystal structure reveals a narrow substrate-binding groove optimized for MHET recognition and hydrolysis.
- **Cutinase** They are naturally occurring enzymes in fungi and bacteria that degrade cutin, a polyester component of plant cuticles. Structurally similar to PETase, many cutinases have demonstrated the ability to hydrolyze PET and other synthetic polyesters. Examples include cutinases from *Thermobifida fusca* and *Humicola insolens*. Their thermostability and broad substrate specificity make them promising candidates for industrial plastic degradation, particularly when used in tandem with PETase.
- **Laccase:** They are multicopper oxidase enzymes that catalyze the oxidation of phenolic and non-phenolic compounds. Although they are not directly involved in hydrolyzing polymer backbones like PET, they can modify the surface chemistry of plastics such as PE and PS, making them more amenable to further enzymatic or microbial attack. Laccases can also degrade plastic additives and dyes, facilitating the detoxification and reuse of plastic waste streams.
- **Lipase:** They catalyze the hydrolysis of ester bonds in lipids but have also shown activity toward synthetic polyesters like PCL and PLA. Their utility lies in degrading biodegradable plastics and potentially in degrading coatings and composite materials. Lipases are commonly derived from *Candida rugosa, Pseudomonas* spp., and *Rhizopus* spp. (Figure 9.1).

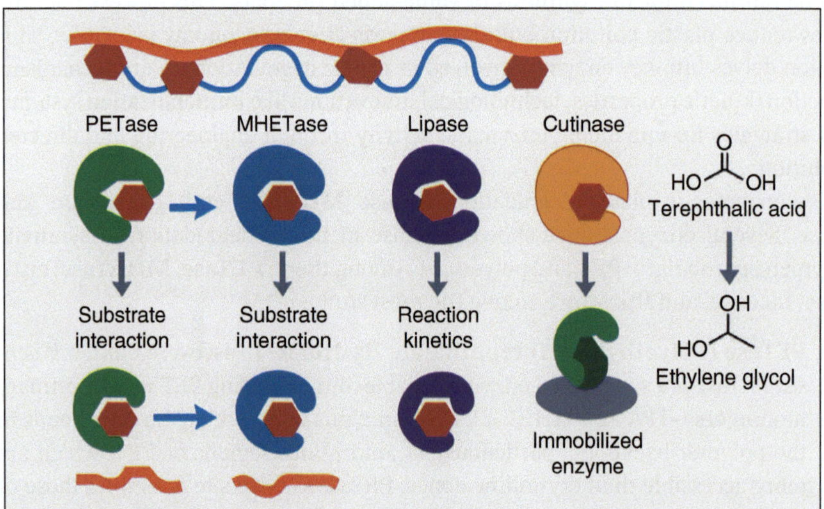

Figure 9.1 Enzymatic degradation of PET plastics via sequential action of PETase, MHETase, lipase, and cutinase to yield terephthalic acid and ethylene glycol.

9.5.1 Enzyme–Substrate Interactions and Reaction Kinetics

Understanding the interaction between enzymes and plastic substrates is vital for optimizing degradation efficiency. These interactions depend on several factors [16]:

- **Polymer Structure:** Enzymes exhibit better binding and activity toward amorphous regions of plastics where polymer chains are loosely packed and more accessible.
- **Surface Area and Hydrophobicity:** Higher surface area improves contact between enzyme and polymer. Enzymes with hydrophobic patches near the active site have enhanced affinity for hydrophobic polymers like PET.
- **Enzyme Catalytic Triad:** Most hydrolytic enzymes (e.g., PETase, cutinase) operate via a serine–histidine–aspartate catalytic triad. This facilitates nucleophilic attack on the ester bond, resulting in bond cleavage and monomer release.
- **Kinetic Parameters (K_m and V_{max}):** The Michaelis–Menten model is often used to quantify enzyme efficiency. A lower K_m indicates higher substrate affinity, while a higher V_{max} indicates a greater maximum rate of reaction. Kinetic profiling helps determine ideal reaction conditions, such as pH, temperature, and substrate concentration.

Temperature, pH, and cofactor presence also play crucial roles in catalytic efficiency. For example, PETase works optimally at around 30–40 °C, but industrial processing may demand higher-temperature stability.

9.5.2 Immobilized Enzyme Systems for Continuous Degradation

While free enzymes offer high specificity, they often suffer from poor stability, difficult recovery, and limited reusability. Immobilization—anchoring enzymes onto solid supports or within matrices—offers a powerful solution to these issues. Several immobilization techniques have been developed [17]:

- **Adsorption onto Carriers:** Enzymes can be adsorbed onto materials like silica, biochar, or magnetic nanoparticles. This method is simple and reversible but may suffer from enzyme leaching.
- **Covalent Binding:** Strong chemical bonds between enzymes and supports offer higher stability but can reduce enzymatic activity due to conformational changes.
- **Entrapment in Gels:** Enzymes can be entrapped within polymers like alginate or polyacrylamide gels, allowing substrate diffusion while protecting the enzyme.
- **Cross-linked Enzyme Aggregates (CLEAs):** Enzymes are aggregated and chemically cross-linked, improving operational stability and resistance to denaturation.

Immobilized enzyme systems are ideal for **continuous flow bioreactors**, where plastic substrates can be fed into the system for constant degradation and product recovery. These systems enable cost-effective and scalable enzymatic recycling, particularly when enzymes with long operational half-lives are used.

9.5.3 Strategies for Improving Enzyme Activity (Mutation, Engineering)

Natural enzymes, while promising, often lack the necessary efficiency, thermostability, or substrate specificity for industrial-scale plastic degradation. Consequently, several strategies have been employed to enhance their properties:

- **Directed Evolution:** This method mimics natural evolution by introducing random mutations in the enzyme's gene and screening for improved variants. For example, evolved variants of PETase have shown significantly higher activity and thermal stability, making them suitable for industrial PET recycling.
- **Site-directed Mutagenesis:** Targeted mutations are introduced at specific amino acid residues believed to influence enzyme activity. Structural insights from X-ray crystallography or molecular dynamics simulations guide this process. For example, mutations in the substrate-binding cleft of PETase have improved its ability to bind and degrade crystalline PET.
- **Protein Engineering:** Rational design using computational tools allows researchers to redesign enzymes with improved properties. This includes modifying surface charge, hydrophobicity, or introducing disulfide bonds to enhance stability.
- **Fusion Proteins:** Enzymes can be fused with binding modules, such as cellulose-binding domains or hydrophobin tags, to improve substrate affinity and surface adsorption. This is especially useful in solid–liquid systems where diffusion is limited.
- **Synthetic Biology Approaches:** Microbial chassis like *Escherichia coli* or *B. subtilis* can be engineered to express plastic-degrading enzymes, enabling whole-cell biocatalysis. Co-expression of PETase and MHETase in a single host system has shown synergistic effects in PET degradation [18].

9.6 Bioconversion of Plastics to Value-added Products

The bioconversion of plastics into value-added products represents a transformative and sustainable approach to mitigating plastic pollution while simultaneously generating economically and environmentally beneficial compounds. One of the well-studied examples is the microbial degradation of PET, a widely used synthetic polymer, into its monomeric constituents TPA and EG. Specific bacteria and enzymes such as PETase and MHETase can efficiently hydrolyze PET into these simpler compounds under optimized conditions. These degradation products, rather than being discarded or incinerated, serve as precursors for various biotechnological applications. For instance, TPA and EG can be further utilized as carbon sources for microbial fermentation to synthesize diverse value-added biochemicals. Among these, biosurfactants stand out due to their biodegradability, surface activity, and applications in bioremediation, cosmetics, and pharmaceuticals. Additionally, microbial pathways can convert plastic-derived substrates into biopolymers such as

PHAs and PLA. PHAs, which are intracellular storage compounds produced by bacteria under nutrient-limited conditions with excess carbon, offer a renewable and biodegradable alternative to petroleum-based plastics. PLA, often synthesized via the microbial fermentation of lactic acid, has gained significant traction in biomedical and packaging industries due to its biocompatibility and thermal processability. Moreover, organic acids such as succinic acid, acetic acid, and adipic acid, which are industrially important platform chemicals, can also be derived from plastic degradation intermediates through engineered microbial pathways. These acids serve as building blocks for bio-based polymers, solvents, and plasticizers, thus creating a closed-loop system of plastic reutilization [19].

Beyond biochemical products, advanced bioconversion strategies focus on upcycling plastic waste into materials for high-value technological applications. For example, the transformation of plastic degradation products into filament precursors for 3D printing has opened new avenues in additive manufacturing. Using microbial or chemo-biological processes, monomers derived from PET or PS can be polymerized into new materials tailored for 3D printing, thus offering a circular economy model where plastic waste becomes the feedstock for precision manufacturing. In parallel, bioconversion processes are also being explored for generating biofuels and fuel additives from plastic waste. Enzymatic hydrolysis followed by microbial fermentation can yield short-chain alcohols, hydrocarbons, or fatty acid derivatives that function as biofuels or additives in gasoline and diesel. These biofuels not only offer a renewable alternative to fossil fuels but also help address the dual challenge of plastic pollution and energy demand. Central to the success of these bioconversion strategies is the use of metabolically engineered microbes tailored for high-efficiency plastic valorization. Through synthetic biology and metabolic engineering, microbial strains are being designed to possess optimized pathways for the assimilation of plastic-derived substrates and the targeted production of desired compounds. For example, *E. coli, P. putida*, and *Cupriavidus necator* have been genetically modified to enhance their tolerance to plastic monomers, increase substrate uptake efficiency, and redirect metabolic fluxes toward the synthesis of PHAs, organic acids, or other industrially relevant metabolites. Furthermore, advanced techniques such as CRISPR-Cas systems, adaptive laboratory evolution, and systems biology are being employed to refine these microbial chassis for robustness, productivity, and scalability. These tailored microbes can be incorporated into bioreactor systems designed to handle mixed plastic waste streams and variable feedstock compositions, thereby improving the overall viability of bioconversion as a commercial-scale solution. By integrating enzymatic depolymerization with microbial biosynthesis, this approach not only diverts plastic waste from landfills and oceans but also contributes to the growing bioeconomy by converting waste into fuels, materials, and chemicals with high market value. Overall, the bioconversion of plastics to value-added products stands at the intersection of environmental sustainability, biotechnology innovation, and circular economic principles, offering a promising strategy for addressing the global plastic crisis [20].

9.7 Synthetic Biology and Metagenomics Approaches

Synthetic biology and metagenomics are emerging as transformative tools in the field of plastic waste valorization, offering innovative solutions to overcome the limitations of traditional biodegradation methods. One of the primary applications of these technologies is genome mining for the discovery of novel plastic-degrading enzymes. In plastic-contaminated ecosystems such as landfills, marine debris zones, or industrial waste sites, microbial communities have naturally evolved to utilize synthetic polymers as carbon sources. Through metagenomic approaches where DNA is directly extracted from environmental samples without the need for culturing, researchers can explore the vast uncultured microbial diversity and identify genes encoding enzymes capable of degrading various plastics such as PE, PET, PS, or PUs. High-throughput sequencing combined with bioinformatic tools enables the annotation of these genes and the prediction of their functional roles. Novel enzymes like PETase and MHETase, originally discovered from *I. sakaiensis*, were identified through such strategies, revealing their ability to hydrolyze PET into monomers like TPA and EG. Genome mining allows scientists to search across thousands of environmental genomes for homologous or functionally analogous enzymes, broadening the catalog of plastic-degrading biocatalysts and paving the way for more efficient biodegradation technologies [21].

In parallel, synthetic biology offers the tools to optimize and engineer these enzymes and their host microbes for enhanced performance. One of the key strategies is pathway engineering, which involves modifying and assembling entire metabolic pathways within microbial hosts to ensure efficient conversion of plastic degradation products into high-value biochemicals. For example, when a plastic polymer like PET is broken down into its monomers, these intermediate products can enter engineered metabolic pathways that lead to the synthesis of bioplastics like PHA, biofuels such as ethanol or butanol, or organic acids including succinic and adipic acids. By rerouting metabolic flux, eliminating competing pathways, and optimizing regulatory elements, pathway engineering significantly improves yield, productivity, and specificity. Microorganisms like *E. coli*, *P. putida*, and *C. necator* are commonly used chassis organisms due to their well-characterized genetics and compatibility with synthetic biology tools. These engineered microbes can be tailored to not only degrade plastics more efficiently but also to synthesize products that meet industrial quality standards.

A crucial aspect of implementing these engineered pathways is the use of plasmid-based expression systems and CRISPR-Cas optimization techniques. Plasmids serve as vectors for introducing foreign genes into microbial hosts, allowing for the controlled expression of plastic-degrading enzymes or entire synthetic metabolic pathways. Modern expression systems are designed to be inducible, tightly regulated, and modular, allowing researchers to fine-tune gene expression based on environmental cues or process needs. Meanwhile, the advent of CRISPR-Cas technology has revolutionized the field of genome editing, enabling precise, efficient, and multiplexed genetic modifications. CRISPR tools can be used to delete undesirable genes, insert beneficial ones, or modulate gene expression levels within the host organism, thereby accelerating the development of robust microbial strains capable of processing plastic waste into valuable end-products. Furthermore, CRISPR interference (CRISPRi) and activation (CRISPRa) systems allow for the dynamic regulation of gene networks without permanently altering the genome, offering flexibility for industrial-scale bioprocesses [22].

Complementing synthetic biology efforts, metagenomic exploration continues to reveal critical insights into the ecology and functionality of plastic-degrading microbial communities. By studying the microbiomes of environments heavily impacted by plastic pollution, researchers gain a deeper understanding of community dynamics, interspecies interactions, and functional gene networks that support plastic degradation. Shotgun metagenomics, for instance, enables the comprehensive profiling of all genetic material within a sample, allowing for the identification of not just individual enzymes but also entire metabolic modules and symbiotic pathways. These insights are particularly valuable for designing synthetic consortia, engineered microbial communities that mimic natural ecosystems but are optimized for stability and productivity in bioreactors. In addition, single-cell genomics and transcriptomic analyses can uncover active degraders and their real-time gene expression patterns in response to different plastic types, enhancing our ability to tailor biodegradation strategies to specific waste streams [23].

Moreover, coupling metagenomics with synthetic biology enables a feedback loop where newly discovered enzymes or genetic modules can be rapidly synthesized, tested, and optimized in lab-based systems. Advances in DNA synthesis, modular cloning, and automated high-throughput screening further accelerate this discovery-to-application pipeline. In some cases, genes encoding plastic-degrading enzymes are expressed in non-native hosts like yeast or cyanobacteria to create photosynthetic or anaerobic systems for plastic valorization, expanding the versatility of biotechnological platforms. Together, synthetic biology and metagenomics represent a powerful convergence of discovery and design. Metagenomics uncovers nature's untapped genetic wealth, while synthetic biology transforms this potential into scalable, efficient, and environmentally compatible solutions. This integrative approach not only enhances the biological degradation of recalcitrant plastic polymers but also facilitates their conversion into a wide array of sustainable products ranging from biopolymers and surfactants to platform chemicals and fuels. As technologies continue to evolve, the synergy between genome mining, pathway optimization, and engineered microbial systems holds the promise of a future where plastic waste is no longer a persistent pollutant but a renewable feedstock in a thriving circular bioeconomy [24].

9.8 Case Studies and Recent Research Breakthroughs

In recent years, significant progress has been made in the development of biological systems for plastic waste valorization. These breakthroughs have moved from lab-scale experimentation to industrial pilots and real-world applications, driven by the urgency to address the growing global plastic pollution crisis. Four key case studies—industrial pilots for PET biodegradation, the discovery of plastic-eating enzymes in marine bacteria, algae–microbe hybrid systems, and innovative solutions developed by biotech startups—highlight the rapid advancements and potential of biological methods to transform plastic waste into valuable resources.

- **Industrial Pilots for PET Biodegradation**
  One of the most promising advancements in plastic waste valorization is the industrial-scale biodegradation of PET, a common plastic used in packaging, textiles, and beverage containers. Companies like Carbios, a French biotech firm, have pioneered enzyme-based recycling technologies that target PET. Carbios developed a proprietary enzymatic process utilizing a thermostable cutinase originally sourced from *T. fusca*, which efficiently depolymerizes PET into its monomers: TPA and EG. This process takes place under moderate temperature and pH conditions and can degrade over 90% of PET in just a few hours. The success of Carbios's technology has led to the establishment of industrial pilot facilities capable of processing tons of postconsumer PET waste. These facilities integrate the enzymatic hydrolysis process with monomer purification and repolymerization, creating a closed-loop recycling system. Unlike traditional mechanical recycling, which often leads to lower-quality plastic products, this biotechnological approach allows for infinite recycling without loss of performance or material quality. The partnership between Carbios and major corporations like PepsiCo and L'Oréal further illustrates the growing industry confidence in enzymatic recycling as a viable and sustainable alternative [25].
- **Plastic-eating Enzymes in Marine Bacteria**
  Another groundbreaking discovery involves the identification of plastic-degrading enzymes in marine bacteria, especially those inhabiting plastic-contaminated ocean environments. In one notable study, researchers isolated a bacterium named *I. sakaiensis* from a PET-contaminated wastewater site in Japan. This bacterium produces two key enzymes, PETase and MHETase, that work in tandem to break down PET into its monomers. The structure of PETase was resolved using X-ray crystallography, and subsequent protein engineering efforts have further improved its efficiency and stability at ambient temperatures.

  Following this discovery, similar enzymes have been identified in various marine microorganisms, particularly those forming biofilms on ocean plastic debris. These microbes have adapted to survive by utilizing plastic as a carbon source, offering insights into natural evolutionary solutions to plastic pollution. Scientists are now using metagenomic techniques to mine marine samples for other potential plastic-degrading genes, expanding the catalog of biocatalysts suitable for environmental and industrial applications. These enzymes, when expressed in host organisms like *E. coli* or *P. putida*, can be used in controlled systems for large-scale biodegradation, thereby offering an eco-friendly solution to marine plastic contamination [26].
- **Algae–Microbe Hybrid Systems for Valorization**
  Hybrid systems involving algae and bacteria represent a novel approach to plastic waste valorization by combining the metabolic capabilities of both organisms. Algae, particularly microalgae such as *Chlorella vulgaris* and *Spirulina*

platensis, are photosynthetic and capable of producing oxygen and fixed carbon, which support the growth of plastic-degrading bacteria. In these systems, bacteria such as *Pseudomonas* or *Bacillus* species degrade plastics like PE or PS into smaller molecules, which are then metabolized by algae for biomass production or converted into valuable biochemicals. These integrated systems have been explored in bioreactor settings, where plastic particles are added to a mixed culture of algae and bacteria. The symbiotic relationship improves system efficiency—bacteria benefit from the oxygen and nutrients released by algae, while algae utilize carbon dioxide and degradation intermediates produced by bacteria. Furthermore, algae can be harvested to extract biofuels, bioplastics, pigments, or feed supplements, adding economic value to the waste treatment process. While still in early stages of development, algae–microbe hybrid systems offer a sustainable, decentralized model for small-scale communities or remote regions dealing with plastic waste challenges [27].

- **Tech Startups and Biotech Solutions Targeting Plastic Waste**
  The growing interest in biological plastic degradation has given rise to numerous startups and biotech ventures focused on commercializing innovative solutions. Companies like Novoloop (USA), BioCellection (now rebranded as Novoloop), and BASF's trinamiX are developing chemical and biological pathways to upcycle plastics into high-value materials. Novoloop, for instance, has developed a process called ATOD™ (Accelerated Thermal Oxidative Decomposition), which breaks down PE into functional chemical intermediates used to produce thermoplastic PUs for footwear and electronics. Another example is Lactips (France), which produces a water-soluble, biodegradable plastic alternative using casein (milk protein). Although not a degradation-based method, their innovation addresses the need for alternatives to single-use plastics and can complement biological degradation strategies. Similarly, Loop Industries employs depolymerization technologies using low-energy inputs to break down PET and polyester fibers into monomers, which are then repolymerized into virgin-quality plastic [28].

  These startups are often supported by venture capital, government grants, and collaborations with academic institutions. They bridge the gap between lab-scale research and scalable industrial application, often integrating synthetic biology, machine learning (ML), and artificial intelligence (AI) for enzyme discovery and process optimization. Their efforts demonstrate that biological approaches to plastic waste management are not just theoretical but are rapidly becoming commercially viable, offering scalable and sustainable alternatives to traditional plastic recycling and disposal methods.

9.9 Challenges and Barriers to Commercialization

The commercialization of biological approaches for plastic waste valorization, while highly promising, faces a number of technical, logistical, and societal challenges that must be addressed before widespread implementation can occur. One of the most critical barriers is the stability, specificity, and production cost of plastic-degrading enzymes. Enzymes like PETase, MHETase, cutinase, and laccase have shown great potential in breaking down synthetic polymers into monomeric units. However, many of these enzymes suffer from poor stability under industrial conditions such as high temperature, varying pH, or the presence of toxic plastic additives. Moreover, their activity is often substrate-specific, meaning that an enzyme optimized for PET may be ineffective against PE or PS. Engineering enzymes to have broad substrate range, high catalytic efficiency, and resilience in harsh conditions is still a work in progress. Additionally, the large-scale production of enzymes through microbial fermentation or synthetic biology remains cost-intensive. The downstream processing, purification, and formulation of these enzymes add to the overall economic burden, making the bio-based processes less competitive compared to conventional plastic recycling or incineration, particularly in regions with limited financial resources or infrastructure.

Another major challenge lies in the integration of bioconversion technologies into existing waste management systems. Most municipal and industrial waste handling frameworks are designed for mechanical or chemical recycling, landfill disposal, or incineration. Introducing a biological valorization process requires a reconfiguration of these systems, including the segregation of plastic types, provision of pretreatment units (e.g., shredding, cleaning, or thermal conditioning), and installation of bioreactors optimized for microbial or enzymatic activity. This poses both financial and logistical hurdles, especially in densely populated urban settings or in developing countries where waste management infrastructure is already under strain. Additionally, biological degradation processes typically require longer residence times than physical or chemical methods, which could lead to bottlenecks in waste throughput unless continuous processing systems are designed and implemented at scale [29].

Contamination and the heterogeneity of plastic waste streams further complicate the commercialization landscape. Real-world plastic waste is rarely composed of a single polymer type; it often includes mixed plastics (e.g., PET with PE or PVC), layered composites, dyes, fillers, additives, and even organic residues such as food or oils. These contaminants not only reduce the efficiency of enzymatic or microbial degradation but may also produce toxic by-products that inhibit microbial metabolism or compromise product quality. Furthermore, some plastic types, such as thermosets or multilayered packaging materials, are inherently resistant to both biological and mechanical degradation. Effective preprocessing techniques such as sorting, washing, and classification become essential, but they increase the cost and complexity of the overall valorization system. Automated systems using AI-driven optical sensors or near-infrared spectroscopy have shown potential for better sorting, but these technologies are still being refined for field use and remain capital-intensive.

Finally, regulatory and public acceptance issues present nontechnical but equally significant barriers to commercialization. Regulatory agencies in different countries have varying standards for waste treatment technologies, product safety, and environmental release of genetically modified organisms (GMOs). The use of engineered microbes or enzymes in open or semi-closed environments raises concerns about biosafety, horizontal gene transfer, and unintended ecological impacts. Stringent regulations and prolonged approval processes can delay the deployment of promising technologies, particularly in regions with cautious or outdated regulatory frameworks. On the public side, skepticism about the safety of synthetic biology and GMOs, along with a lack of awareness about the benefits of biological valorization, can lead to resistance from communities and stakeholders. Consumer perception also plays a crucial role in the market viability of products derived from plastic waste. Without adequate communication, transparency, and education about the environmental and economic benefits of such technologies, public resistance could slow down adoption.

In conclusion, while biological valorization of plastic waste holds immense potential to address the global plastic crisis in a sustainable and circular manner, several challenges must be overcome to transition from laboratory success to commercial viability. These include the technical limitations of current enzymes and microbes, the need for infrastructure adaptation, the complexity of waste streams, and regulatory as well as social hurdles. Overcoming these barriers will require coordinated efforts in research, policy, public engagement, and industrial collaboration to ensure that biological plastic valorization becomes not only feasible but also economically and socially sustainable on a global scale [30].

9.10 Future Perspectives and Research Directions

As the urgency to combat plastic pollution intensifies, future perspectives and research directions in biological plastic valorization are shifting toward more integrated, intelligent, and scalable solutions. One of the foremost areas of advancement lies in the discovery of new microbial strains and enzymes capable of degrading diverse plastic polymers. Current enzymatic tools like PETase, MHETase, and laccase have shown promise, but the plastic-degrading potential of the global microbial community remains largely untapped. Environmental microbiomes from extreme habitats such as landfills, deep-sea plastic debris, hot springs, and oil-contaminated soils are likely to harbor novel organisms with unique metabolic capabilities. Through techniques like metagenomics, single-cell sequencing, and functional screening, researchers aim to identify and characterize new enzymes that are more robust, efficient, and active under industrial conditions. Expanding the catalog of plastic-degrading organisms will enable the tailoring of microbial consortia to target specific polymer types, enhance biodegradation rates, and reduce the environmental footprint of treatment processes. Alongside natural discovery, AI and ML are emerging as powerful tools for optimizing enzyme function and pathway design. By analyzing vast datasets of protein structures, substrate binding affinities, and kinetic parameters, AI algorithms can predict mutations that improve enzyme stability, specificity, or catalytic activity. ML models are also being used to simulate entire metabolic networks, allowing researchers to design and fine-tune synthetic pathways for converting plastic-derived monomers into high-value products like bioplastics, organic acids, or fuels. This computational approach significantly accelerates the enzyme engineering cycle, reducing the trial-and-error nature of traditional laboratory work. Integration of AI in synthetic biology workflows will enhance the predictability, efficiency, and scalability of biological systems designed for plastic valorization.

Another critical research frontier is the development of advanced bioreactors for continuous plastic degradation. Traditional batch systems are often inefficient for large-scale waste processing due to limitations in substrate loading, oxygen transfer, and microbial activity. Future designs are focusing on modular, scalable, and automated bioreactors that support continuous feeding of plastic waste, real-time monitoring of degradation products, and sustained microbial or enzymatic activity. Innovations such as immobilized enzyme systems, biofilm-based reactors, and membrane bioreactors are being explored to improve degradation efficiency, reduce operating costs, and facilitate downstream product recovery. These systems can be optimized for specific types of plastic waste and engineered to operate under various environmental conditions, making them suitable for both centralized facilities and small-scale applications.

Finally, the integration of biological plastic degradation into decentralized and urban waste management models presents a transformative opportunity for localized, circular solutions. Urban areas generate vast amounts of plastic waste, often mixed with organic matter and other contaminants. Embedding compact, biologically based treatment units in neighborhoods, industries, or commercial centers can enable on-site conversion of plastic into useful products like biogas, compostable polymers, or chemical feedstocks. Such decentralized systems not only reduce transportation and landfill costs but also empower communities to participate in sustainable waste management. Coupled with smart monitoring, automation, and community education, these models support resilient, scalable, and adaptive solutions to the plastic crisis.

In summary, the future of plastic waste valorization lies in a multidisciplinary approach that combines microbial discovery, computational design, bioprocess engineering, and systems integration. Continued innovation and collaboration across these domains will be essential to realize the full potential of biological technologies in building a cleaner, circular, and more sustainable planet.

9.11 Conclusion

Biological valorization of plastic waste offers a paradigm shift in addressing one of the most pressing environmental challenges of our time. Leveraging nature's enzymatic machinery and microbial versatility, plastics once considered indestructible can now be transformed into valuable resources. Though technological and economic barriers remain, interdisciplinary research, synthetic biology, and bioprocess engineering are rapidly advancing the feasibility of such approaches. A future where biological systems play a central role in the plastic life cycle is not only plausible but also necessary for realizing a sustainable, circular economy.

References

1 Yoshida, S., Hiraga, K., Takehana, T., et al. (2016). A bacterium that degrades and assimilates poly(ethylene terephthalate). *Science*. 351 (6278): 1196–1199.

2 Tanasupawat, S., Takehana, T., Yoshida, S., et al. (2016). Characterization of the PET-degrading bacterium *Ideonella sakaiensis. Int J Syst Evol Microbiol*. 66 (10): 3591–3595.

3 Wei, R. and Zimmermann, W. (2017). Biocatalysis as a green method for polyester waste recycling. *Trends Biotechnol*. 35 (6): 563–572.

4 Austin, H.P., Allen, M.D., Donohoe, B.S., et al. (2018). Characterization and engineering of a plastic-degrading aromatic polyesterase. *Proc Natl Acad Sci USA*. 115 (19): E4350–E4357.

5 Danso, D., Chow, J., and Streit, W.R. (2019). Plastics: environmental and biotechnological perspectives on microbial degradation. *Appl Environ Microbiol*. 85 (19): e01095–e01119.

6 Urbanek, A.K., Rymowicz, W., and Mirończuk, A.M. (2018). Degradation of plastics and plastic-degrading bacteria in cold marine habitats. *Appl Microbiol Biotechnol*. 102 (18): 7669–7678.

7 Ru, J., Huo, Y., and Yang, Y. (2020). Microbial degradation and valorization of plastic wastes. *Front Microbiol*. 11: 442.

8 Kawai, F., Kawabata, T., and Oda, M. (2020). Current knowledge on enzymatic PET degradation and its possible application to waste stream management and other fields. *Appl Microbiol Biotechnol*. 104 (20): 7715–7732.

9 Sudhakar, M., Doble, M., Murthy, P.S., and Venkatesan, R. (2008). Marine microbe-mediated biodegradation of low- and high-density polyethylenes. *Int Biodeterior Biodegradation*. 61 (3): 203–213.

10 Ghosh, S.K., Pal, S., and Ray, S. (2013). Study of microbes having potentiality for biodegradation of plastics. *Environ Sci Pollut Res*. 20 (7): 4339–4355.

11 Shah, A.A., Hasan, F., Hameed, A., and Ahmed, S. (2008). Biological degradation of plastics: a comprehensive review. *Biotechnol Adv*. 26 (3): 246–265.

12 Auta, H.S., Emenike, C.U., and Fauziah, S.H. (2017). Distribution and importance of microplastics in the marine environment: a review of the sources, fate, effects, and potential solutions. *Environ Int*. 102: 165–176.

13 Syranidou, E., Karkanorachaki, K., Amorotti, F., et al. (2019). Biodegradation of mixture of plastic films by tailored marine consortia. *Sci Total Environ*. 697: 134128.

14 Mohanan, N., Montazer, Z., Sharma, P.K., and Levin, D.B. (2020). Microbial and enzymatic degradation of synthetic plastics. *Front Microbiol*. 11: 580709.

15 Nanda, S. and Berruti, F. (2021). Municipal solid waste management and landfilling technologies: a review. *Environ Chem Lett*. 19 (2): 1433–1453.

16 Nakajima-Kambe, T., Onuma, F., Kimpara, N., and Nakahara, T. (1995). Isolation and characterization of a bacterium which utilizes polyester polyurethane as a sole carbon and nitrogen source. *FEMS Microbiol Lett*. 129 (1): 39–42.

17 Pathak, V.M. (2017). Review on the current status of polymer degradation: a microbial approach. *Bioresour Bioprocess*. 4 (1):1–31.

18 Webb, H.K., Arnott, J., Crawford, R.J., and Ivanova, E.P. (2013). Plastic degradation and its environmental implications with special reference to poly(ethylene terephthalate). *Polymers.* 5 (1): 1–18.

19 Yousif, E. and Haddad, R. (2013). Photodegradation and photostabilization of polymers, especially polystyrene: review. *SpringerPlus.* 2 (1): 398.

20 Chamas, A., Moon, H., Zheng, J., et al. (2020). Degradation rates of plastics in the environment. *ACS Sustain Chem Eng.* 8 (9): 3494–3511.

21 Kroon, F.J., Motti, C.A., Talbot, S., et al. (2018). A workflow for improving estimates of microplastic contamination in marine waters: a case study from the Great Barrier Reef. *Mar Pollut Bull.* 128: 579–586.

22 Ammendolia, J., Saturno, J., Brooks, B., et al. (2021). An emerging source of plastic pollution: environmental presence of plastic personal protective equipment (PPE) debris related to COVID-19 in a metropolitan city. *Environ Pollut.* 269: 116160.

23 Tiwari, R.R., Srivastava, J.K., and Shukla, S.K. (2020). Biodegradation of plastics: an overview. *J Bioremediat Biodegrad.* 11: 466.

24 Ali, S.S., Elsamahy, T., Al-Tohamy, R., et al. (2021). Plastic wastes biodegradation: mechanisms, challenges and future prospects. *Sci Total Environ.* 780: 146590.

25 Khatami, K., Pishgar, R., Mehrabadi, H., et al. (2021). Waste-to-energy technologies for municipal solid waste management: a review. *Chemosphere.* 277: 130885.

26 Singh, B. and Sharma, N. (2008). Mechanistic implications of plastic degradation. *Polym Degrad Stab.* 93 (3): 561–584.

27 Ragaert, K., Delva, L., and Van Geem, K. (2017). Mechanical and chemical recycling of solid plastic waste. *Waste Manag.* 69: 24–58.

28 Singh, J., Gupta, K., and Singh, S. (2022). Emerging trends in enzymatic plastic degradation: recent advances and future prospects. *Environ Sci Pollut Res.* 29: 14694–14714.

29 Llorca, M., Schirinzi, G.F., Martínez, M., et al. (2018). Adsorption of perfluoroalkyl substances on microplastics under environmental conditions. *Environ Pollut.* 235: 680–688.

30 Wierckx, N., Prieto, A., Pomposiello, P., et al. (2015). Plastic waste as a novel substrate for industrial biotechnology. *Microb Biotechnol.* 8 (6): 900–903.

10

Biological Approaches to the Valorization of Nuclear Wastes

10.1 Introduction

Nuclear energy has long been regarded as a powerful and efficient source of electricity generation, capable of meeting large-scale energy demands with relatively low greenhouse gas emissions. Since the mid-twentieth century, countries around the world have harnessed nuclear power for civilian and military applications, contributing significantly to global electricity production. As of recent estimates, over 440 operational nuclear reactors exist worldwide, providing approximately 10% of the world's electricity. While the benefits of nuclear energy, such as low operational carbon footprint, high energy density, and reduced dependence on fossil fuels, are considerable, the technology comes with a critical downside: the generation of radioactive wastes.

Radioactive waste is an inevitable by-product of nuclear fission reactions, nuclear fuel reprocessing, research activities, and medical and industrial uses of radioactive isotopes. This waste remains hazardous for varying periods, depending on its composition, and requires stringent handling, storage, and disposal protocols. Managing radioactive waste poses one of the most significant challenges to the sustainable operation of the nuclear industry, primarily due to its long-lived radiotoxicity and environmental risks. Nuclear wastes are classified based on their radioactivity levels, heat generation, and potential environmental hazard. The most widely used classification includes low-level waste (LLW), intermediate-level waste (ILW), and high-level waste (HLW). **LLW** includes items like contaminated clothing, tools, filters, and other materials with relatively low levels of radioactivity. These wastes typically originate from hospitals, research institutions, and nuclear power plants and do not require shielding during handling and transport. They are usually disposed of in near-surface landfill facilities. **ILW** contains higher amounts of radioactivity and may require shielding, although it does not generate significant heat. ILW includes resins, chemical sludges, and reactor components. Disposal of ILW often involves more engineered containment and deeper burial. **HLW** is the most hazardous category and includes used nuclear fuel and waste from the reprocessing of spent fuel. HLW generates substantial

Biological Processes and Techniques in Zero Waste Management, First Edition. Garima Awasthi and Mohit Nigam. Published 2026 by Wiley-VCH GmbH.

heat and remains radioactive for thousands to millions of years. Safe, long-term management of HLW typically involves deep geological repositories, which are costly and technically complex to develop. The improper management or accidental release of nuclear waste can have devastating environmental and human health consequences. Legacy contamination from past nuclear weapons testing, accidents (e.g., Chernobyl, Fukushima), and poorly regulated waste disposal has led to persistent radioactive pollution in various ecosystems. Radionuclides like cesium-137, strontium-90, uranium, plutonium, and technetium-99 can migrate through soil and groundwater, enter food chains, and pose long-term risks to both wildlife and human populations [1].

In addition to direct radiological hazards, nuclear waste sites often face issues such as chemical toxicity, corrosion of storage containers, and challenges in long-term monitoring. The decommissioning of aging nuclear facilities and remediation of contaminated sites further complicate waste management efforts. Moreover, public opposition to nuclear activities is often driven by concerns over environmental contamination and the unresolved problem of HLW disposal. In response to the limitations of conventional nuclear waste treatment and disposal methods, biological approaches have emerged as innovative, eco-friendly alternatives for mitigating radiological and chemical hazards. These biological strategies leverage the capabilities of microorganisms, plants, and enzymes to interact with radionuclides and transform them into less toxic or immobilized forms. Processes such as **bioremediation**, **biosorption**, **bioaccumulation**, and **biotransformation** are at the forefront of this field. Microorganisms, including bacteria, fungi, and algae, can adapt to highly radioactive environments and perform redox transformations that alter the solubility and mobility of radionuclides. Certain microbes produce extracellular polymeric substances (EPSs) or biofilms that bind radionuclides, effectively reducing their bioavailability and migration. Similarly, plants used in **phytoremediation** can uptake radionuclides from soil or water and sequester them in their tissues, facilitating easier containment and disposal. Advances in **synthetic biology** and **genetic engineering** are further enhancing the capabilities of these biological agents, enabling the development of engineered microbes with tailored functions for specific radionuclide targets. These technologies offer the promise of *in situ* waste treatment, cost-effectiveness, and reduced environmental impact compared to physical and chemical methods. Biological valorization also introduces the possibility of recovering valuable isotopes or converting waste into usable forms, aligning with the principles of circular economy and sustainable resource use. Although still in the early stages of development and application, biological approaches are increasingly gaining attention as complementary or alternative strategies to conventional nuclear waste management [2].

10.2 The Need for Biological Valorization of Nuclear Waste

The management of nuclear waste has traditionally relied on physical and chemical treatment methods such as solidification, vitrification, cementation, ion exchange, and chemical precipitation. While these methods can be effective in stabilizing or isolating radioactive materials, they often fall short in terms of long-term sustainability, cost-efficiency, and environmental safety. Many of these techniques require high energy inputs, generate secondary waste streams, and involve complex engineering infrastructure, making them both economically and ecologically burdensome. Furthermore, they do not fundamentally eliminate the radiotoxicity of the waste but rather focus on containment, which merely postpones the environmental risk. Long-term storage, especially for HLW, remains a critical challenge. Deep geological repositories, while technically sound in theory, are fraught with uncertainties related to geological stability, potential groundwater intrusion, and long-term monitoring. These concerns are compounded by strong public resistance to nuclear waste disposal sites, often driven by fears of radiation leakage, health risks, and environmental contamination. Social acceptability thus becomes a major barrier to the deployment of large-scale nuclear waste management infrastructure.

In light of these limitations, biological valorization presents a compelling alternative. Unlike conventional methods, biological approaches operate under mild environmental conditions, are self-sustaining once established, and can be implemented *in situ*, reducing the need for transportation and centralized facilities. Microorganisms and plants can immobilize, transform, or extract radionuclides through processes such as bioreduction, biosorption, and phytoremediation, offering potential for long-term stabilization of contaminants with minimal ecological disturbance. Moreover, the development of genetically engineered microbes and enzymes tailored for specific radionuclides holds promise for highly selective and efficient waste treatment. These biological strategies offer not only environmental compatibility but also align closely with the principles of the **circular economy** by enabling the recovery of useful isotopes or materials from waste streams. By converting waste into resources and minimizing environmental hazards, biological valorization supports broader goals of **low-carbon development**, ecological stewardship, and sustainable industrial practice. As the global community seeks resilient and green technologies to address mounting environmental and energy challenges, integrating biological solutions into the nuclear waste management framework represents a forward-looking and socially responsible step. Continued research, policy support, and public engagement will be essential to realizing the full potential of these emerging biotechnological interventions [3].

10.3 Categories and Characteristics of Nuclear Wastes

Nuclear wastes, generated from power generation, nuclear weapons programs, medical diagnostics, and research activities, exist in various physical states, solid, liquid, and gaseous, each presenting distinct challenges for management and remediation. Solid radioactive wastes include spent nuclear fuel rods, contaminated equipment, filters, and protective clothing. These are generally more manageable due to their contained form but often require shielding and long-term storage due to high levels of radioactivity. Liquid wastes, originating from reactor cooling systems, reprocessing plants, and laboratory operations, pose a higher risk of environmental contamination, as they can infiltrate soil and groundwater systems. These liquids often contain dissolved radionuclides and require complex treatment processes, such as ion exchange or evaporation, before disposal. Gaseous radioactive wastes, including isotopes of iodine, krypton, and tritium, are particularly challenging due to their volatility and mobility in the atmosphere. They can be released during fuel reprocessing or reactor venting and may pose inhalation risks if not properly contained and filtered.

A critical aspect of nuclear waste management lies in understanding the chemical and radiological behavior of specific key radionuclides present in these wastes. Some of the most environmentally significant and hazardous radionuclides include Uranium (U), Plutonium (Pu), Technetium (Tc), Cesium (Cs), and Strontium (Sr). Uranium and plutonium, both actinides, are alpha-emitting elements with long half-lives and are highly toxic even in trace amounts. Plutonium-239, for instance, has a half-life of over 24,000 years and poses serious health risks if inhaled or ingested. Technetium-99, a fission product with a half-life of approximately 211,000 years, is notable for its high mobility in the environment, particularly in its pertechnetate form (TcO_4^-), which is soluble and difficult to immobilize. Cesium-137 and strontium-90 are beta emitters with half-lives of around 30 years; they mimic potassium and calcium, respectively, making them biologically available and potentially harmful when incorporated into living tissues. These radionuclides can persist in the environment, bioaccumulate in organisms, and contribute to long-term radiological hazards. Understanding the half-lives, toxicity, and migration potential of these radionuclides is essential for designing effective waste treatment and bioremediation strategies. Radionuclides with longer half-lives pose enduring threats to ecosystems, especially when they are chemically stable and soluble in water, enabling migration through soil and aquifers. Their toxicity is influenced not only by their radiation type and intensity but also by their chemical behavior in biological systems. The potential for radionuclides to move through food chains and accumulate in organisms adds complexity to environmental risk assessments. Central to predicting radionuclide behavior is the speciation—the chemical form in which an element exists under given environmental conditions. Speciation directly influences mobility, bioavailability, and reactivity, making it a critical parameter in nuclear waste bioremediation. For instance, uranium can exist in several oxidation states (U(IV), U(VI)), with U(VI) being more soluble and mobile than the reduced U(IV) form, which tends to precipitate and immobilize. Microbial and geochemical processes can alter radionuclide speciation, thereby enhancing or limiting their transport. Effective bioremediation strategies aim to transform radionuclides into less soluble or biologically unavailable forms through microbial metabolism or bio-induced precipitation. Understanding the redox potential, pH, presence of complexing agents, and microbial communities in contaminated environments is crucial for influencing these transformations. In summary, a comprehensive understanding of the categories, radionuclide properties, and environmental behaviors of nuclear waste is essential for developing safe and effective remediation strategies.

The complex interplay of physical state, radionuclide chemistry, and ecological dynamics underscores the importance of targeted, science-driven approaches—especially biological ones—for managing and valorizing nuclear waste. As research in microbial and plant-based remediation evolves, integrating knowledge of radionuclide speciation and migration potential will be instrumental in mitigating the long-term impacts of radioactive contamination [4].

10.4 Microbial Systems in Nuclear Waste Transformation

10.4.1 Extremophiles in Radiation Environments

Extremophilic microorganisms, particularly those adapted to survive in highly radioactive environments, play a pivotal role in the biological transformation and stabilization of nuclear waste. Among the most studied radiation-resistant extremophiles is *Deinococcus radiodurans*, often referred to as the world's toughest bacterium. This organism can withstand ionizing radiation doses exceeding 5,000 Gy, far beyond the lethal limit for humans. Its extraordinary resistance stems from a combination of robust DNA repair mechanisms, including efficient homologous recombination, nucleotide excision repair, and a unique ability to reassemble shattered chromosomes with high fidelity. Additionally, *D. radiodurans* produces high levels of antioxidative enzymes such as superoxide dismutase and catalase, which neutralize reactive oxygen species generated by radiation-induced oxidative stress. These enzymes, along with metal-chelating molecules like manganese complexes, protect cellular proteins and DNA from oxidative damage. Another notable extremophile is *Thermococcus gammatolerans*, an archaeon isolated from deep-sea hydrothermal vents, which thrives at high temperatures and can tolerate gamma radiation doses up to 30 kGy. Its ability to survive in extreme conditions is attributed to the presence of highly stable biomolecules, efficient DNA repair systems, and protective cellular structures. Like *D. radiodurans, T. gammatolerans* exhibits an enhanced antioxidative defense system that minimizes radiation-induced damage and supports its survival in radioactive and thermally extreme niches. Other extremophiles, such as *Rubrobacter radiotolerans, Halobacterium salinarum*, and *Arthrobacter* species, also contribute to radionuclide transformation through processes like biosorption, biomineralization, and enzymatic redox reactions [5].

The survival capabilities and metabolic versatility of these organisms make them promising candidates for **bioremediation of radioactive environments**. Their resilience allows them to remain active in high-radiation settings where most microbial life would perish, making them ideal for applications such as *in situ* biotransformation of radionuclides or bio-barrier formation around contaminated zones. By harnessing their natural resistance mechanisms and potentially enhancing them through genetic engineering, extremophilic microbes offer a sustainable, biologically driven approach to mitigating the risks associated with long-lived radioactive waste.

10.4.2 Metal-reducing and Sulphate-reducing Bacteria

Metal-reducing and sulfate-reducing bacteria play a crucial role in the biogeochemical cycling of metals and radionuclides, particularly in the context of nuclear waste transformation and environmental remediation. Among the most extensively studied genera in this group are ***Geobacter***, ***Shewanella***, and ***Desulfovibrio***, which possess unique metabolic capabilities that allow them to interact with and transform radionuclides through redox reactions. These bacteria use metals and radionuclides as terminal electron acceptors during anaerobic respiration, a process that can lead to the immobilization and detoxification of otherwise mobile and hazardous radioactive elements.

Geobacter species are particularly known for their ability to reduce soluble **U(VI)** (uranyl ion) to insoluble **U(IV)** (uraninite), thereby precipitating uranium out of groundwater and preventing its migration. This redox transformation significantly reduces the bioavailability and environmental mobility of uranium. *Geobacter* also participates in the reduction of other radionuclides such as technetium (Tc) and neptunium (Np), making it a key organism in bioremediation strategies for contaminated nuclear sites. Its electron transport chain involves c-type cytochromes and conductive pili (nanowires) that facilitate extracellular electron transfer to insoluble metal oxides [6].

Shewanella species demonstrate remarkable versatility in reducing a wide range of metal and radionuclide species. *S. oneidensis*, for example, can reduce Fe(III), Mn(IV), Cr(VI), and U(VI) under anaerobic conditions, often through the action of outer membrane cytochromes such as MtrC and OmcA. These proteins enable electron transfer across the cell envelope to extracellular mineral surfaces, leading to the reduction and immobilization of radionuclides. *Shewanella's* flexibility in utilizing different electron donors and acceptors makes it highly adaptable to fluctuating environmental conditions in nuclear waste sites.

Desulfovibrio species, belonging to the sulfate-reducing bacterial group, contribute to radionuclide transformation by generating sulfide as a metabolic by-product, which can precipitate radionuclides as insoluble sulfide minerals. Additionally, *Desulfovibrio* can directly reduce radionuclides like U(VI) and Tc(VII) through enzymatic pathways involving hydrogenases and cytochromes. This dual capability, both direct enzymatic reduction and indirect chemical transformation via sulfide production, enhances their effectiveness in immobilizing radioactive contaminants in subsurface environments. Collectively, these bacteria serve as powerful tools for *in situ* bioremediation by transforming soluble, mobile radionuclides into stable, insoluble forms. Their natural redox capabilities, coupled with adaptability to anaerobic conditions, offer promising avenues for the development of biologically based containment and stabilization strategies in radioactive waste management.

10.4.3 Archaea and Fungi in Radioactive Environments

While bacteria are well-known for their role in radionuclide transformation, archaea and fungi also demonstrate significant potential in radioactive waste environments due to their unique physiological traits and adaptive capabilities. Both groups contribute to the stabilization, immobilization, and detoxification of radionuclides through mechanisms such as biosorption and biomineralization, especially in extreme or heavily contaminated conditions where other microorganisms may not survive. Archaea, particularly those belonging to the extremophilic domains such as halophiles, thermophiles, and acidophiles, are remarkably well-adapted to harsh environments, including those contaminated with radioactive waste. These microorganisms exhibit high resistance to metal toxicity and ionizing radiation, largely due to their robust cell envelopes, efficient DNA repair systems, and specialized proteins that stabilize cellular components under stress. Archaea can perform biosorption, in which radionuclides adhere to their cell walls via interactions with functional groups such as carboxyl, phosphate, and hydroxyl moieties. In some cases, they also induce biomineralization, converting soluble radionuclides into insoluble mineral forms like phosphates or carbonates, thereby reducing their mobility and bioavailability. For example, haloarchaea found in saline waste disposal sites have been shown to bind uranium and cesium ions, offering potential for their use in bioremediation of saline radioactive environments.

Fungi, especially filamentous types such as *Aspergillus, Penicillium*, and *Trichoderma*, also exhibit strong potential in radioactive environments due to their exceptional metal tolerance and high surface area for interaction with contaminants. Fungal cell walls, rich in chitin, glucans, and proteins, offer numerous binding sites for radionuclides, facilitating biosorption processes. In addition, fungi are capable of bioprecipitation, a form of biomineralization where they alter the pH or redox conditions of their environment to promote the formation of insoluble radionuclide complexes. Fungi also produce extracellular enzymes and organic acids that can chelate metals or alter their chemical speciation, enhancing the immobilization of radionuclides such as uranium, strontium, and cesium. Many fungi are known for their resistance to ionizing radiation, an adaptation thought to be linked to the presence of protective pigments like melanin, which can absorb and dissipate radiation energy. Together, archaea and fungi broaden the scope of biological remediation by operating under extreme conditions and employing diverse biochemical strategies. Their ability to tolerate and transform radionuclides positions them as valuable allies in biotechnological solutions for nuclear waste management, particularly in environments unsuitable for bacterial activity [7].

10.5 Biological Processes and Mechanisms

Biological systems offer a versatile array of mechanisms for transforming, immobilizing, or detoxifying radioactive contaminants in the environment. These processes often involve microbial or fungal activity and are governed by complex biochemical and physicochemical interactions. The most significant mechanisms involved in the biological remediation of nuclear waste include biosorption, bioaccumulation, bioprecipitation, biomineralization, enzymatic reduction, microbially induced redox transformations, and biofilm-mediated immobilization. Each of these processes contributes uniquely to the stabilization and valorization of radionuclides in contaminated environments.

- **Biosorption and Bioaccumulation**
  Biosorption is a passive physicochemical process by which radionuclides adhere to the surface of biological materials such as microbial cell walls, fungal hyphae, or algal biomass. This binding typically occurs through ionic interactions, chelation, complexation, and van der Waals forces between the radionuclide ions and functional groups like carboxyl, hydroxyl, amino, and phosphate groups on the cell surface. One key advantage of biosorption is that it does not require metabolically active cells, making it an efficient and low-cost method for rapid radionuclide removal from aqueous solutions.

  Bioaccumulation, by contrast, is an active, metabolism-dependent process in which living cells absorb radionuclides into their intracellular compartments. This uptake often involves specific transport proteins or channels and may include intracellular sequestration or compartmentalization to minimize cellular damage. While slower than biosorption, bioaccumulation allows cells to store radionuclides over longer periods and in some cases, convert them into less toxic or insoluble forms. Both processes are valuable in designing biosorbent materials and *in situ* remediation systems using microbial biomass [8].
- **Bioprecipitation and Biomineralization**
  Bioprecipitation involves the conversion of soluble radionuclides into insoluble precipitates through microbially mediated changes in the local chemical environment. Microorganisms can alter pH and redox potential or release metabolic by-products (e.g., sulfides, phosphates, carbonates) that react with radionuclides, forming stable, insoluble compounds. For example, sulfate-reducing bacteria can produce hydrogen sulfide, which reacts with metal ions such as uranium or technetium to form insoluble metal sulfides, thus removing the contaminants from solution.

  Closely related is **biomineralization**, a process by which living organisms induce the formation of mineral phases either within or outside their cells. Biomineralization may be biologically controlled (genetically regulated) or biologically induced (resulting from metabolic by-products). In radionuclide-contaminated environments, this can lead to the formation of uranium–phosphate minerals, calcium-uranium complexes, or other mineralized forms that immobilize radionuclides in a stable matrix. Biomineralization not only stabilizes radionuclides but also contributes to the long-term geochemical isolation of radioactive contaminants.

- **Enzymatic Reduction of Radionuclides**
  Microorganisms possess enzymes capable of catalyzing the **reduction of radionuclides** from one oxidation state to another, typically converting soluble forms into insoluble ones. This process is crucial in the detoxification and immobilization of various radioactive elements. For instance, certain metal-reducing bacteria can enzymatically reduce soluble **U(VI)** to **U(IV)**, which precipitates as insoluble uraninite (UO_2), thereby limiting uranium's mobility in groundwater. Key enzymes involved in these reactions include **hydrogenases**, **cytochromes**, and **reductases**, which transfer electrons from organic or inorganic donors to the target radionuclide. Enzymatic reduction is highly specific and can occur under anaerobic or microaerophilic conditions, depending on the organism. It is particularly significant for redox-sensitive radionuclides such as uranium, technetium, chromium, and neptunium. The efficiency of enzymatic reduction depends on various factors, including the availability of electron donors, environmental pH, and the presence of competing electron acceptors such as nitrate or sulfate [9].
- **Microbially Induced Redox Transformation**
  Beyond enzymatic reactions, microorganisms can influence the **redox chemistry** of radionuclides through broader metabolic activity that alters environmental redox conditions. These **microbially induced redox transformations** can involve both direct electron transfer to radionuclides and indirect effects via the reduction of metals and other electron acceptors that, in turn, influence radionuclide speciation. For example, the reduction of iron(III) to iron(II) by iron-reducing bacteria can result in the co-reduction or coprecipitation of radionuclides such as uranium, plutonium, or technetium. Such transformations often lead to the conversion of radionuclides into less soluble oxidation states, which are more easily immobilized. This microbial activity plays a key role in the natural attenuation of radionuclide contamination in sediments and aquifers, and is increasingly being harnessed in engineered bioremediation systems.
- **Biofilm-mediated Containment and Immobilization**
  Biofilms—structured communities of microorganisms embedded in EPSs—offer a stable matrix for the immobilization of radionuclides. The EPS matrix, composed of polysaccharides, proteins, lipids, and nucleic acids, provides a high density of functional groups that can bind radionuclides through biosorption. Moreover, biofilms offer microenvironments with steep redox and pH gradients, which can enhance precipitation and redox transformations within the film structure. Biofilms also provide resistance to radiation, toxic metals, and desiccation, allowing microbial communities to survive and function in contaminated environments over extended periods. Their protective nature and surface-binding capacity make them particularly useful in creating **bio-barriers**, where biofilms are cultivated on permeable reactive barriers (PRBs) or subsurface interfaces to trap and immobilize radionuclides as groundwater passes through (Figure 10.1) [10].

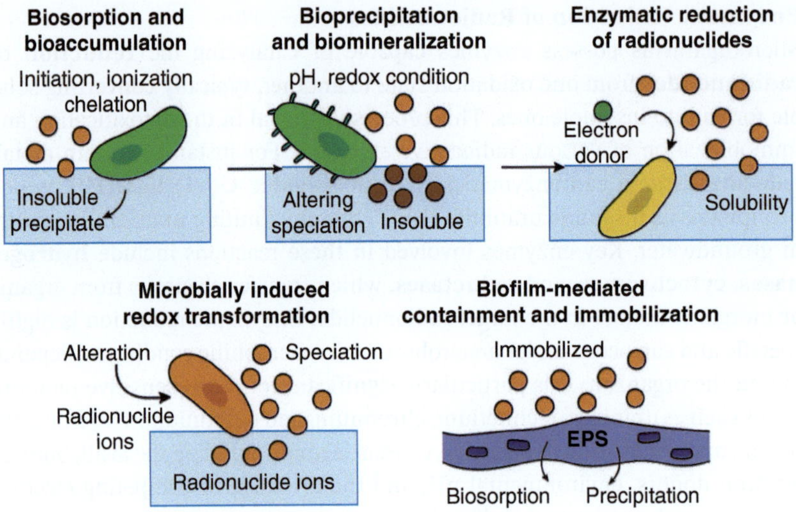

Figure 10.1 Biological mechanisms for radionuclide transformation and immobilization.

10.6 Synthetic Biology and Genetic Engineering for Enhanced Performance

The growing concern surrounding nuclear waste contamination has driven the need for more efficient, precise, and sustainable strategies for radioactive waste remediation. Traditional physicochemical methods often fall short in terms of cost, scalability, or environmental compatibility. Biological approaches, especially those rooted in synthetic biology and genetic engineering, have emerged as promising alternatives. These modern techniques allow the reprogramming of organisms to perform complex tasks, such as metal uptake, transformation of radionuclides, and real-time environmental sensing. Three key advancements in this area include the engineering of *D. radiodurans* and *Shewanella* spp. for enhanced metal uptake, the use of CRISPR tools for pathway optimization, and the development of biosensors for real-time biotransformation monitoring.

10.6.1 Engineering *D. radiodurans* and *Shewanella* spp. for Higher Metal

Deinococcus radiodurans is a well-known extremophilic bacterium famous for its extraordinary resistance to ionizing radiation and desiccation. This makes it an ideal chassis organism for bioremediation in highly radioactive environments. However, its natural capacity to uptake or immobilize heavy metals and radionuclides is limited. Through synthetic biology, its genome can be edited to express specific metal-binding peptides, transporters, or metabolic enzymes, thereby boosting its bioremediation efficiency.

One successful strategy involves the overexpression of metal ion transport proteins such as P-type ATPases, which facilitate the uptake and accumulation of ions like uranium (U), cesium (Cs), and cobalt (Co). Additionally, genes encoding metallothioneins or phytochelatins—small peptides with high affinity for metal ions—can be introduced into *D. radiodurans* to sequester radioactive contaminants intracellularly. Moreover, pathways involved in biosorption and bioprecipitation can be optimized to increase the retention of radionuclides on the bacterial surface or within EPSs.

Similarly, *S. oneidensis*, a facultative anaerobe with versatile metabolic capabilities, is another prime candidate for bioremediation, particularly in subsurface and anoxic environments. Its ability to reduce metals like uranium (VI) to uranium (IV), thereby precipitating it out of solution, is especially valuable. Genetic engineering efforts in *Shewanella* have focused on enhancing electron transport chains, outer membrane cytochromes (e.g., MtrC, OmcA), and flavin secretion pathways to boost metal reduction rates. Synthetic constructs can also be designed to couple the metal-reducing activity of *Shewanella* with metal-binding tags or nanoparticle conjugation to facilitate efficient capture and removal. Together, these engineered microorganisms form a powerful toolkit for site-specific bioremediation of radioactive waste, offering advantages such as targeted uptake, resistance to harsh conditions, and enhanced degradation capabilities [11].

10.6.2 CRISPR Tools for Pathway Enhancement

Clustered Regularly Interspaced Short Palindromic Repeats (CRISPR)-based systems have revolutionized genetic engineering by enabling precise, programmable edits to microbial genomes. In the context of radioactive waste bioremediation, CRISPR tools allow the fine-tuning of metabolic and regulatory pathways to enhance microbial performance.

CRISPR-Cas9 and its derivatives (e.g., CRISPR-Cas12, Cas13, dCas9) can be employed to knock in or knock out specific genes involved in radionuclide transport, resistance, or transformation. For instance, introducing regulatory circuits that overexpress key genes only in the presence of radiation or specific metal ions can reduce metabolic burden on the host and improve fitness in contaminated environments.

In *D. radiodurans*, CRISPR can be used to construct synthetic operons that link radiation-resistance genes with metal uptake pathways, allowing the bacterium to simultaneously tolerate and process radioactive materials more effectively. In *Shewanella*, multiplexed CRISPR approaches can be used to rewire entire respiratory pathways, thereby increasing electron flux toward metal-reducing processes. Additionally, CRISPR interference (CRISPRi) systems can be deployed to repress genes that divert metabolic flux away from desired remediation pathways. Beyond simple gene editing, CRISPR-based logic circuits and programmable gene networks can be constructed to enable intelligent, context-dependent behavior. For example, cells could be designed to activate biotransformation only when multiple environmental cues are present, such as both a high level of radiation and the presence of a target radionuclide, thus ensuring safety and specificity. The modularity and efficiency of CRISPR make it a cornerstone technology in designing next-generation bioremediation systems capable of addressing the complexity and variability of radioactive waste sites [12].

10.6.3 Biosensors for Real-time Monitoring of Radioactive Biotransformation

Monitoring the progress and effectiveness of bioremediation processes in real-time is critical for environmental safety, process control, and regulatory compliance. Genetically encoded biosensors provide a powerful solution to this challenge by translating biological activity into measurable signals, such as fluorescence, luminescence, or color change. Biosensors can be engineered using promoter-reporter gene fusions, where the expression of a reporter (e.g., GFP, luciferase, or β-galactosidase) is driven by promoters that respond to specific stimuli, such as metal ions (e.g., U, Cs, Sr) or radiation-induced stress. These sensors can be integrated into *D. radiodurans* or *Shewanella*, enabling on-site, real-time visualization of biotransformation activity. For example, a uranyl-sensitive biosensor might use a synthetic promoter responsive to uranium uptake, coupled with a fluorescence output. Upon detecting uranium, the engineered cells emit a signal, indicating successful sequestration. This allows operators to assess bioremediation progress without the need for destructive sampling or complex instrumentation.

Advanced biosensors can also incorporate logic-gated systems that respond to multiple environmental signals, enhancing specificity. For instance, a biosensor might only activate in response to both pH change and metal presence, reducing false positives. Moreover, biosensors can be coupled with wireless data transmission modules or integrated into microfluidic devices for remote or high-throughput monitoring. In addition to sensing contaminants, biosensors can monitor cellular health, metabolic flux, or oxidative stress levels, parameters critical for maintaining the robustness of microbial systems in extreme environments. The integration of biosensors with engineered pathways also allows for real-time feedback control, enabling dynamic adjustment of gene expression to optimize remediation outcomes [13].

10.7 Valorization Outcomes and Resource Recovery

The biological remediation of radioactive waste not only serves to mitigate environmental contamination but also presents valuable opportunities for the recovery and reuse of critical materials. Through carefully engineered microbial processes and innovative biotechnological strategies, radioactive waste can be transformed from a hazardous burden into a source of economically and strategically important elements. Valorization in this context refers to the transformation of waste into valuable products. This approach aligns with principles of the circular economy and sustainable resource management. The three major outcomes of valorization in radioactive bioremediation include the **recovery of rare earth elements (REEs) and isotopes**, **conversion of radioactive ions into stable mineral forms**, and **reuse of recovered materials in medical, industrial, or technological applications**.

10.7.1 Recovery of REEs and Isotopes

REEs and certain isotopes found in radioactive waste are of immense strategic importance due to their critical role in high-tech industries, electronics, renewable energy systems, and national defense. However, their natural reserves are limited, and conventional extraction methods are costly and environmentally damaging. Interestingly, many of these elements, including lanthanides and actinides, are present in spent nuclear fuel, mining tailings, and other radioactive waste streams.

Microbial systems, particularly those that have been genetically modified or adapted for metal tolerance, can be used to selectively bind, accumulate, or precipitate REEs. Biosorption using microbial cell walls or EPSs is one mechanism through which REEs can be concentrated. Certain bacteria can be engineered to express peptides or metalloproteins with high specificity for rare earth ions, allowing selective recovery even from complex mixtures.

For example, lanthanides such as neodymium and europium, critical for magnets, lasers, and electronic displays, can be isolated using bio-based chelating agents. In parallel, valuable radioactive isotopes like cobalt-60, iodine-131, or molybdenum-99, which are used in medical imaging and cancer therapy, can also be recovered using microbial biosorption or precipitation. The use of biologically assisted ion exchange or biocolloids for selective capture offers a scalable and sustainable pathway to reclaim these valuable resources [14].

10.7.2 Transformation of Radioactive Ions into Stable Mineral Forms

A major challenge in the management of radioactive waste is the long-term containment of radionuclides. Many radioactive ions, such as uranium (U), technetium (Tc), strontium (Sr), and cesium (Cs), are highly mobile in groundwater and pose significant risks to ecosystems and human health. Through microbial processes, these ions can be converted into **insoluble, stable mineral forms** that are less likely to leach into the environment.

This transformation is achieved via **bioprecipitation, biomineralization, or enzymatic reduction** mechanisms. For example, *S. oneidensis* and *Geobacter sulfurreducens* are known to reduce soluble U(VI) to insoluble U(IV), which precipitates as uraninite (UO_2). Similarly, radioactive strontium and barium can be coprecipitated with microbial sulfate to form barite ($BaSO_4$), which effectively traps Sr^{2+} within its crystal lattice.

Microbially induced carbonate precipitation (MICP) is another promising pathway where radionuclides like thorium or plutonium are immobilized within carbonate matrices. Such biologically stabilized forms exhibit high durability and resistance to leaching, making them suitable for long-term containment. Genetic engineering and synthetic biology approaches can further enhance these pathways by overexpressing key enzymes or improving the metabolic pathways responsible for redox and precipitation reactions.

This biological stabilization strategy not only reduces environmental risk but also simplifies waste storage logistics and lowers the long-term costs associated with radioactive waste management [15].

10.7.3 Reuse in Medical Isotopes, Radiation Shielding, or Catalysts

Once recovered and stabilized, many materials from radioactive waste can be **repurposed for high-value applications**. This reuse not only adds economic value but also reduces the demand for virgin materials and the environmental footprint of mining or refining processes.

One of the most important applications is in the **medical field**, where specific isotopes such as cobalt-60, iodine-131, and lutetium-177 are crucial for cancer radiotherapy, imaging (e.g., PET and SPECT scans), and diagnostic procedures. Isotopes reclaimed from bioprocessed waste can meet this demand with high purity if properly separated and sterilized. This also ensures a more reliable and cost-effective supply of medical isotopes, which are often in short supply due to complex production logistics.

Recovered radioactive materials can also be used in **radiation shielding**, especially in industrial or nuclear settings. Lead and depleted uranium, when safely immobilized, can be fashioned into shielding materials for nuclear reactors, radiological labs, and even space exploration applications. The use of biologically extracted and stabilized materials reduces reliance on environmentally destructive mining.

In addition, several radionuclides and metal ions, such as cerium, thorium, and vanadium, serve as **catalysts** in industrial chemical reactions. These can be reused in refining processes, polymerization, or environmental cleanup operations. Through biosynthetic mineral formation or nanoparticle conjugation, these elements can be recovered in forms that are suitable for industrial application, thus closing the loop in a sustainable manner [16].

10.8 Case Studies and Applied Research

The application of biological approaches for radioactive waste management has evolved from theoretical concepts to real-world implementation through various field trials and research case studies. These efforts illustrate how microbial processes can be effectively used to transform, immobilize, or contain radionuclides in diverse environmental contexts. Several pioneering projects around the world have tested these concepts under field conditions or pilot-scale setups, generating valuable insights into the feasibility, efficiency, and scalability of bioremediation. Prominent examples include uranium bioremediation at contaminated mining sites, pilot-scale bioreactor systems, natural attenuation processes in aquifers, and integrated strategies combining biology with geochemical and engineering controls.

One of the most well-documented and influential case studies in microbial uranium remediation comes from the Oak Ridge Reservation in Tennessee, USA. This site, historically associated with nuclear weapons production and uranium enrichment, became heavily contaminated with uranium and other heavy metals, particularly in the subsurface and groundwater environments.

In the early 2000s, the US Department of Energy (DOE) launched a large-scale bioremediation project under its Subsurface Biogeochemical Research (SBR) program. The objective was to evaluate whether indigenous or introduced microorganisms could reduce soluble uranium (U(VI)) to insoluble uranium (U(IV)), thereby immobilizing it within the subsurface. The biostimulation strategy employed involved injecting organic carbon sources such as ethanol or acetate into the subsurface to stimulate the activity of native metal-reducing bacteria, particularly *Geobacter* species [17].

Field studies at the Oak Ridge site showed promising results: uranium concentrations in groundwater were significantly reduced as microbial communities flourished and converted U(VI) to U(IV), forming stable uraninite (UO_2) precipitates. Moreover, molecular and geochemical monitoring revealed long-term persistence of the reduced uranium under favorable redox conditions. This case provided concrete proof that *in situ* bioremediation using native microbial populations is not only feasible but can also be scaled to address complex contamination at nuclear legacy sites.

In addition to *in situ* field applications, controlled **pilot-scale bioreactor systems** have been developed to treat low-level radioactive waste (LLRW) under engineered conditions. These systems allow for optimization of microbial activity, control over reaction parameters, and containment of radioactive materials, making them suitable for treatment of effluents from hospitals, laboratories, or decommissioning operations.

One notable example includes fluidized-bed bioreactors designed to treat effluents containing radionuclides like strontium-90 (Sr-90), cesium-137 (Cs-137), and cobalt-60 (Co-60). These reactors are often inoculated with metal-tolerant or genetically engineered bacteria capable of biosorption, bioaccumulation, or enzymatic transformation of radionuclides. Nutrient delivery, pH, temperature, and retention time are carefully regulated to maintain optimal microbial performance.

In some designs, the reactors incorporate **immobilized biomass or biofilms** on carriers, which improves stability and prevents washout. The treated effluent often meets regulatory discharge standards, while the biomass containing immobilized radionuclides can be safely separated and disposed of. These pilot-scale studies serve as testbeds for developing full-scale, modular systems suitable for on-site treatment and resource recovery, especially in remote or mobile settings.

Natural attenuation refers to the passive reduction of contaminant concentrations through naturally occurring physical, chemical, or biological processes. In radioactive waste management, this approach has been studied in aquifers contaminated by long-term seepage from nuclear facilities or tailing ponds. Although it does not involve direct human intervention, **monitoring natural attenuation** helps understand subsurface microbial ecology and informs risk assessments and containment strategies.

For example, in aquifers affected by uranium mining, microbial communities capable of reducing U(VI) to U(IV) may already exist. These indigenous populations, often anaerobic iron- or sulfate-reducing bacteria, can naturally immobilize radionuclides under the right redox conditions. Over time, contaminants become stabilized as mineral precipitates or adsorbed onto sediments, reducing their mobility and bioavailability.

Extensive hydrogeological, geochemical, and microbiological monitoring is essential to evaluate the effectiveness of natural attenuation. At sites like Hanford (Washington, USA) and Ranger (Australia), long-term studies have shown that microbial populations can play a central role in radionuclide cycling and attenuation. However, reliance on natural attenuation alone may not always be sufficient due to changing environmental conditions, and it is typically used as part of a monitored natural attenuation (MNA) strategy supported by contingency plans. While microbial processes are powerful tools for radionuclide immobilization, their effectiveness can be significantly enhanced when **integrated with geochemical and physical containment strategies**. This multidisciplinary approach leverages the synergy between biological, chemical, and engineering interventions to achieve more robust and long-lasting remediation outcomes.

One common integration involves coupling microbial reduction with geochemical amendments such as **phosphate or sulfate addition**. These amendments promote the formation of stable mineral phases like metal phosphates or sulfides, which further trap radionuclides. For instance, in uranium-contaminated sites, the combined use of phosphate and microbial activity can lead to the precipitation of uranium-phosphate minerals with very low solubility, ensuring long-term containment.

Physical containment measures, such as PRBs, subsurface slurry walls, or engineered clay liners, can be used to isolate contaminated zones while microbial processes operate within or behind these barriers. In some designs, bioactive PRBs are created by injecting nutrients or bacteria into the barrier zone to create a "living filter" that degrades or immobilizes contaminants as groundwater passes through.

Additionally, integrating real-time **biosensor technology** and geochemical monitoring helps track bioremediation progress and ensure safety. Advanced modeling tools are also employed to simulate subsurface transport, microbial kinetics, and long-term behavior of radionuclides under varying environmental scenarios.

This integrated approach, combining biology with geochemistry and engineering, represents the most promising and scalable strategy for remediating complex and heterogeneous radioactive waste sites [18].

10.9 Challenges, Risks, and Ethical Considerations

While biological approaches to radioactive waste remediation offer promising advantages in terms of sustainability, cost-effectiveness, and environmental compatibility, they are not without significant challenges, risks, and ethical implications. One of the foremost technical limitations is the issue of **incomplete degradation and residual toxicity**. Although microbial systems can transform or immobilize certain radionuclides, complete detoxification is not always achieved. Some intermediate products may retain hazardous properties or become remobilized under fluctuating environmental conditions, such as changes in pH, temperature, or redox potential. For instance, while uranium may be reduced to its less soluble form (U(IV)), oxidative disturbances in the environment could reoxidize it back to a mobile and toxic form (U(VI)). Similarly, biosorption processes may temporarily bind contaminants to microbial surfaces or biofilms, but these associations are sometimes reversible, particularly under high ionic strength or competitive conditions. This poses a long-term risk of recontamination, especially in subsurface environments where monitoring is difficult and slow biochemical changes may go unnoticed.

Another significant concern revolves around **biosafety, particularly in handling radioactive biomass**. Microorganisms used for radionuclide remediation, whether native, adapted, or genetically modified, accumulate radioactive elements during the bioremediation process. As a result, the biomass itself becomes hazardous and must be treated, stored, or disposed of as radioactive waste. This creates a secondary waste management challenge. Moreover, improper handling, disposal, or containment of this biomass could lead to the release of radioactive particles into the environment or occupational exposure for workers. The risk is compounded when genetically engineered microorganisms are used, as their behavior in natural ecosystems can be unpredictable. If released unintentionally, they may alter microbial communities, outcompete native species, or transfer engineered genes through horizontal gene transfer, potentially resulting in ecological imbalances or unforeseen consequences.

In addition to biosafety, there are substantial **regulatory and societal hurdles** that must be addressed. Regulatory frameworks for radioactive waste management are traditionally designed around physicochemical treatment methods and are often ill-equipped to accommodate or evaluate biological technologies. Existing safety standards, licensing procedures, and environmental compliance assessments may not fully capture the complexity and uniqueness of microbial remediation. Furthermore, the deployment of genetically engineered organisms in open environments is subject to strict biosafety regulations and often requires extensive environmental impact assessments, which can delay or hinder field applications. Public perception also plays a critical role in the acceptance of these technologies. The dual use of "radioactive" and "genetically modified" terms in the same context often triggers public concern, skepticism, or outright opposition. Misinformation or lack of awareness about how these systems work can generate resistance from local communities or stakeholders, even in cases where risks are minimal and well-managed.

Finally, **long-term monitoring and risk assessment** represent essential but complex aspects of any bioremediation strategy. Unlike conventional methods where waste is physically removed or contained in engineered structures, biological remediation relies on dynamic processes that are influenced by multiple environmental variables. This necessitates continuous and often multi-decadal monitoring of the remediation site to ensure stability, containment, and lack of recontamination. Advanced molecular tools, biosensors, and geochemical modeling can assist in monitoring, but they require technical expertise, funding, and consistent oversight.

Moreover, risk assessments must account for worst-case scenarios, including the potential reactivation of radionuclides, failure of containment barriers, or loss of microbial activity over time. Ethical questions also arise in terms of intergenerational responsibility: decisions made today regarding biological remediation may have implications for future populations, particularly if monitoring systems fail or legacy sites are inadequately managed. While biological remediation holds immense promise in addressing radioactive waste challenges, it must be pursued with caution, supported by rigorous scientific validation, strong regulatory oversight, effective risk communication, and ethical responsibility. Addressing these concerns transparently and proactively will be crucial to ensuring that these innovative approaches are not only technically successful but also socially and environmentally responsible.

10.10 Conclusion

Biological valorization of nuclear waste is a frontier domain that converges microbiology, biotechnology, and nuclear science to address one of the most pressing environmental concerns of the modern era. Microbial systems offer a sustainable, adaptive, and potentially low-cost approach to manage and transform radioactive wastes, especially when integrated with modern synthetic biology and environmental engineering. While the road to full-scale deployment is fraught with technical, ethical, and regulatory hurdles, the potential of biological methods to contribute meaningfully to nuclear waste remediation and resource recovery is undeniable. Investment in research, safety frameworks, and cross-sector collaboration will be key to realizing the full promise of this transformative approach.

References

1 Butterworth, S.J., Barton, F., and Lloyd, J.R. (2023). Extremophilic microbial metabolism and radioactive waste disposal. *Extremophiles*. 27 (3): 27.
2 Wang, S., Wang, J., Tian, Y., and Wang, J. (2024). Uranium removal in groundwater by *Priestia* sp. isolated from uranium-contaminated mining soil. *Chemosphere*. 351: 141204.
3 Yu, X., Xiong, F., Zhou, C., et al. (2024). Uranium bioprecipitation mediated by a phosphate-solubilizing *Enterobacter* sp. N1-10 and remediation of uranium-contaminated soil. *Sci Total Environ*. 906: 167688.
4 Chen, Z., Li, Q., Yang, Y., et al. (2022). Uranium removal from a radioactive contaminated soil by defined bioleaching bacteria. *J Radioanal Nucl Chem*. 331 (1): 439–449.
5 Sadowska-Bartosz, I. and Bartosz, G. (2023). Antioxidant defense of *Deinococcus radiodurans*: how does it contribute to extreme radiation resistance? *Int J Radiat Biol*. 99 (12): 1803–1829.

6 Kaksonen, A.H., Lakaniemi, A.M., and Tuovinen, O.H. (2020). Acid and ferric sulfate bioleaching of uranium ores: a review. *J Clean Prod*. 264: 121586.

7 Manobala, T., Shukla, S.K., Rao, T.S., and Kumar, M.D. (2021). Kinetic modelling of the uranium biosorption by *Deinococcus radiodurans* biofilm. *Chemosphere*. 269: 128722.

8 Kulkarni, S., Ballal, A., and Apte, S.K. (2013). Bioprecipitation of uranium from alkaline waste solutions using recombinant *Deinococcus radiodurans*. *J Hazard Mater*. 262: 853–861.

9 Li, L., Jiang, F., Wang, L., et al. (2025). Research on the mechanism of uranium pollution caused by *SUP* gene transformation of *Deinococcus radiodurans*. *J Radioanal Nucl Chem*. 334 (2): 1595–1611.

10 Ivanov, N.P., Dran'kov, A.N., Shichalin, O.O., et al. (2024). Composite magnetic sorbents based on magnetic Fe_3O_4 coated by Zn and Al layered double hydroxide for U(VI) removal from aqueous media. *J Radioanal Nucl Chem*. 333 (3): 1213–1230.

11 Nie, X., Dong, F., Liu, M., et al. (2021). Bonding behavior and mechanism of U(VI) by chemically modified *Deinococcus radiodurans*. *Minerals*. 11 (10): 1108.

12 Kumari, M., Karn, S.K., and Raj, V. (2024). Extremophiles and related extremozymes: their structure-function relationship in industrial applications. *Indus Biotechnol*. 20 (6): 279–295.

13 Fredrickson, J.K., Kostandarithes, H.M., Li, S.W., et al. (2000). Reduction of Fe(III), Cr(VI), U(VI), and Tc(VII) by *Deinococcus radiodurans* R1. *Appl Environ Microbiol*. 66 (5): 2006–2011.

14 Morales Hidalgo, M. (2025). *Influencia de los microorganismos y parámetros fisicoquímicos (radiación, temperatura) en la integridad de las barreras de contenedor de cobre y bentonita de los almacenamientos geológicos profundos*. Doctoral dissertation. Universidad de Granada.

15 Sakai, M., Shimosaka, T., Katsumata, K., et al. (2024). Developing a new host-vector system for *Deinococcus grandis*. *Front Microbiol*. 15: 1387296.

16 Zhang, Z., Tang, Z., Liu, Y., et al. (2023). Study on the ecotoxic effects of uranium and heavy metal elements in soils of a uranium mining area in northern Guangdong. *Toxics*. 11 (2): 97.

17 Ighalo, J.O., Chen, Z., Ohoro, C.R., et al. (2024). A review of remediation technologies for uranium-contaminated water. *Chemosphere*. 352: 141322.

18 Qin, W., Xiao, Y., Zhang, C., et al. (2024). Bismuth nanocluster loaded on N-doped porous carbon with "memory catalysis" effect for efficient photocatalytic H_2 generation and uranium (VI) reduction. *Chem Eng J*. 490: 151511.

Part III

Advanced Tools and Techniques to Achieve Zero Waste

11

In Silico Tool and Techniques Supporting Waste Treatment

11.1 Introduction

Wastes generated are the biggest problem of environment. Society is generating and releasing waste in the environment through various sources. These generated waste materials belong to different chemical groups. These chemicals are of different chemical structures, molecular weights, compositions and concentrations. These chemical substances are present in high concentrations in nature and responsible for polluting the environment. The chemicals in waste materials mostly originate from polynuclear aromatic hydrocarbons used in several anthropogenic activities. These chemicals are in the form of pesticides, herbicides, insecticides, agrochemicals, carcinogens, drugs, dyes, detergents, chemical agents, etc. The huge concentration and the synthetic structural features are the major reason that these chemicals are accepted or not accepted by the environment [1, 2]. These highly toxic organic compounds are released into the environment due to various human activities and get mixed up in the ecosystem. As they are widely applicable in industries and agriculture, often they are responsible for the pollution and contamination of the ecosystem. Such toxic compounds can seriously damage the food chains of the ecosystems, as they affect almost all the trophic levels and can cause many problems [3]. The chemicals generated through waste pose a great threat to the sustainability of our environment. These chemicals are chlorinated dibenzodioxins, polychlorinated biphenyls, halocarbons, hydrocarbons, alkyl benzyl sulphonates, and various oil mixtures. The synthetic chemicals, including metals, are not easily degraded by native flora as compared to pesticides, dyes, drugs, and chemicals, etc. [4].

11.1.1 Chemicals Nature Present in Waste Material

Polycyclic aromatic hydrocarbons (PAH) released mainly from industrial production are the major source of hydrophobic organic contaminants. Restoration of polluted sites can be easily achieved by using technology like biodegradation to reduce

Biological Processes and Techniques in Zero Waste Management, First Edition. Garima Awasthi and Mohit Nigam. Published 2026 by Wiley-VCH GmbH.

or remove PAH with the help of microorganisms. The polychlorinated biphenyls are mostly inert. They are used in industrial and commercial applications in the mixtures of synthetic organic chemicals. These compounds can cause cancer and can act as endocrine disruptors [5].

Nitroaromatic compounds (NACs) are important industrial feedstocks, extensively used as precursors for amino aromatic derivatives, as solvents, and in the synthesis of dyes, plasticizers, pharmaceuticals, pesticides (parathion, methyl parathion, dinoseb, dinitrocresol, and nitrofen), and explosives 2,4,6-trinitrotoluene (TNT), hexahydro-1,3,5-trinitro-1,3,5,7-triazine (RDX), and octahydro-1,3,5,7-tetranitro-1,3,5,7-tetrazocine (HMX) [6]. They are stable, persistent, and toxic when released into the environment. They are highly reactive in mammalians, so they are potentially more carcinogenic or mutagenic. They convert nitro groups into more harmful nitroso and hydroxylamine groups [7].

Pesticides are those substances that are meant to act against pests. They can be persistent or nonpersistent. Commonly pesticides are used as a food source by fungi and bacteria [8, 9].

Hydrocarbons are organic compounds and consist of hydrogen and carbon. These are classified into aromatic and aliphatic compounds. The aliphatic compounds are linear and branched, whereas the aromatic compounds are ring compounds such as benzene [10].

Dyes are used widely in many industries, including textiles, paper, pharmaceuticals, etc. Aromatic compounds with one or more (—N=N—) groups are known as azo dyes. Azo dyes are the largest class of synthetic dyes [11, 12].

Heavy metals can only be stabilized or removed, which can only be achieved through biotransformation. The microbes help in reducing heavy metals through mechanisms such as bioleaching, biosorption, biomineralization, intracellular accumulation, and transformation, which are catalyzed by enzymes [13, 14].

The biodegradability of these compounds is very poor, mainly because of their structures, and due to this reason, the microbes fail to recognize them as a substrate. Secondly, the large benzene-based aromatic structure of waste materials also enables them to enter the microbial cell. Some microbes have been seen to be capable of breaking down waste materials to some extent [12].

11.1.2 Need of Advancement in Microbial Study for Waste Degradation

Alexander proposed in 1965 that the microorganisms are capable to degrade naturally occurring compounds, and the principle is called microbial infallibility [15]. Microorganisms can degrade various waste materials, and they cannot degrade various other such compounds. The compounds that resist biodegradation and persist in the environment are called recalcitrant. The recalcitrant property of waste material is directly linked to its complexity; higher the complexity, stronger the recalcitrant property. Most of the waste materials are recalcitrant, i.e., they cannot be biodegraded [16].

Non-degradation and slow degradation of these materials is a global issue of concern. The average degradation time of these waste materials varies from days to millions of years (Table 11.1) [17].

Table 11.1 Waste materials are their degradation time frame.

S. No.	Waste materials	Estimated decomposition time
1	Fruit and vegetable peels	5 days to 1 month
2	Train ticket paper	Around 2 weeks
3	Regular paper	2–6 months
4	Cotton gloves	Approximately 3 months
5	Wax-coated milk cartons	About 3 months
6	Cotton T-shirt	Around 6 months
7	Orange peels	Roughly 6 months
8	Corrugated cardboard	Around 2 months
9	Cotton or synthetic thread	3–4 months
10	Natural fiber ropes	3 months to over 1 year
11	Canvas-based products	Up to 1 year
12	Tree leaves	Around 1 year
13	Plywood	1–3 years
14	Woolen socks and clothing	1–5 years
15	Plastic-coated paper cartons	Around 5 years
16	Tetra Pak (composite milk containers)	Approximately 5 years
17	Chewing gum	Over 5 years
18	Cigarette butts	1–10 years
19	Tin cans	50 years
20	Aluminum cans	80–200 years
21	Glass bottles	Over 1 million years (does not decompose naturally)
22	Styrofoam (polystyrene foam)	Does not biodegrade
23	Plastic bottles	450–1,000 years
24	Disposable diapers	Around 500 years
25	Fishing line (nylon)	600 years or more
26	Plastic bags (grocery-type)	10–1,000 years (depends on type)
27	Leather shoes	25–40 years
28	Rubber soles	50–80 years
29	Bamboo toothbrush	3–6 months (handle)
30	Biodegradable plastic (PLA, PHA)	3–6 months (under composting)

PLA, polylactic acid; PHA, polyhydroxyalkanoates.

11.1.3 *In Silico* Tools and Techniques for Waste Treatment

The computational simulations and modeling are transforming the field of waste treatment by providing predictive insights, optimizing processes, and guiding experimental design through *in silico* approaches. These tools reduce costs and time and improve the efficiency and sustainability of biological, chemical, and physical waste treatment technologies.

The new ways are in search to get sustainable development to clean up contaminated environments of the society, as the pollution is increasing in the environment day by day. Different research activities are continuously being conducted, which are oriented toward the development of new techniques to fight against these environmental problems. Sustainability of our environment is very important, as the toxic waste materials that are not accepted by our ecosystem and require the process of biodegradation or bioremediation [18]. To degrade the waste before moving to *in vitro* techniques, *in silico* tools and techniques can be used to analyze various properties of waste, microbes, and the interaction of waste against the microbes. The few tools and techniques are shown in Figure 11.1.

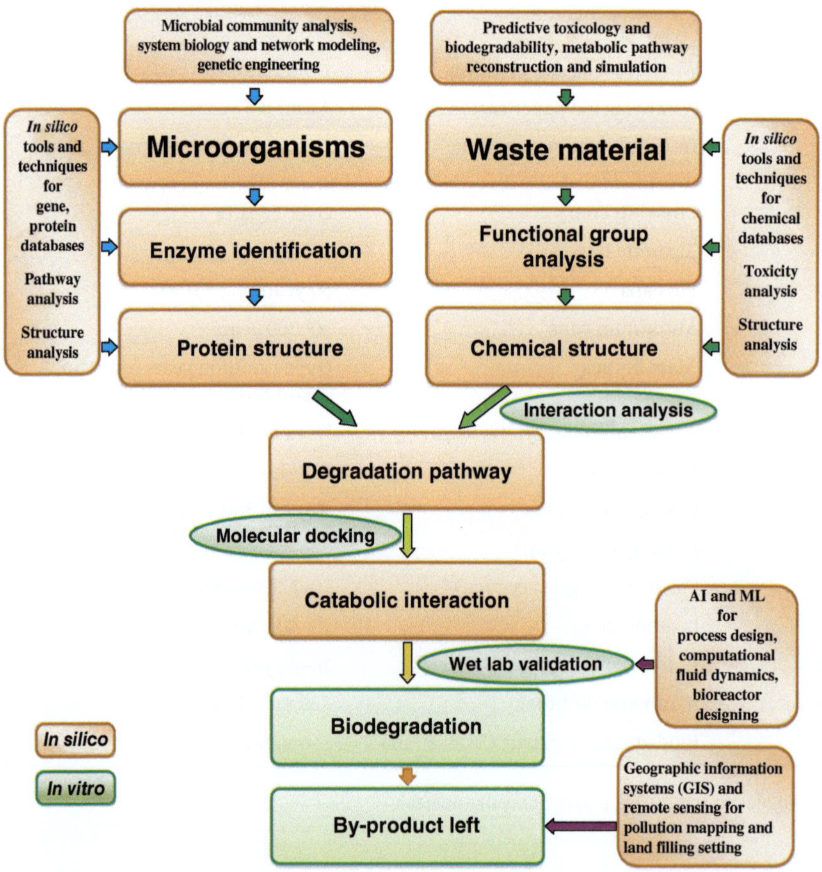

Figure 11.1 Flow diagram showing the application of *in silico* tools and techniques for waste treatment and its association with *in vitro* analysis.

11.2 *In Silico* Tools and Techniques for Data Collection

11.2.1 Literature Survey

The extensive literature searches are done through research papers, review papers, and articles through abstracting from journals available. The identified genes and proteins involved in the degradation of waste material were collected from the gene pool available in literature. The toxicity and various other properties of chemicals present in waste material can be found through already published papers and articles in literature.

11.2.2 *In Silico* Data Mining

11.2.2.1 Chemical Databases

Databases are also available for chemicals, which are an important target for biodegradation study. Various computational approaches are available for information predictions regarding the chemicals, such as toxicity prediction, prediction of physical and chemical properties, functional groups, and degree of degradation process [19]. Chemical databases available are PubChem, ChemIDplus, and ECHA. Mostly the databases provide information such as name, synonyms, chemical formula, and molecular weight, the image of the chemical structure, canonical 3D structure in PDB format, SMILES code, density, melting point, boiling point, evaporation rate, water solubility and links to other related databases.

Few more tools, such as TerraQSAR and SAMFA, predict the molecular properties based on the structure except for online tools [20]. The *in silico* approaches can also be used to predict the toxicity levels of chemicals [21]. The databases of chemicals for toxicity index are ACuteTox, SuperToxic, Terra-Base, GENE-TOX, Aggregated Computational Toxicology Resource (ACToR), Chemical Effects in Biological Systems (CEBS), Chemical Carcinogenesis Research Information System (CCRIS), Comparative Toxicogenomics Database (CTD), Carcinogenic Potency Database (CPDB), Development and Reproductive Toxicology Database (DART), Genetic Toxicity Data Bank (GENETOX), Hazardous Substances Data Bank (HSDB), and Toxicity Literature Online (TOXLINE).

11.2.2.2 Gene and Protein Databases

Data mining was done by NCBI, DDBJ, and EMBL genomic databank and from UM-BBD. The information gathered about gene was its degrading microorganism, accession number, gene sequence in the form of FASTA sequence, and functional protein. The gene mining was done for different waste materials like pesticides, dyes, aromatic compounds, biodegradable plastics, and drugs. Additional information about microbes can be found through various available databases.

To biologically degrade these chemical substances, the study of microbes is parallel important. Various computational tools help in data mining and provide an approach to explore thousands of microbes, their genomic databases, proteomic databases such as ExPASy, NCBI, DDBJ, EMBL, and metabolic pathways databases KEGG, and Bionemo—Molecular information on biodegradation metabolism. It helps in analyzing environmentally relevant microorganisms for gene expression analysis, functional enzyme analysis, metabolic pathway analysis, etc. The genomics study of microbes includes sequence assembly, gene prediction, and bacterial annotation [22].

11.2.3 Homology Search of Gene and Protein

Various tools such as BLAST tool, ALIGN Query, LALIGN, FFAS, FASTA, GeneWise, SIM, and SSEA were used for homology search to identify similar genes. During the search, various genes were identified for the degradation of all the selected waste material. Few genes were similar but reported for different strains of the microorganism, whereas many different genes were also identified during the search. In the second screening, the unique genes were selected based on different genes from total identified genes responsible for degrading waste material. To use BLAST software, the first BLAST website will be opened, and then the entry of accession number or FASTA format of gene sequence will be done. Finally, the homologous gene was identified, and during this search, many more genes were identified, which were capable of degrading waste material.

The protein of selected unique gene was further studied to identify its superfamily (conserved regions). This protein was identified using BLASTp and ScanProsite tool. The functional classification of waste-degrading genes was done based on the motif and its function. The entry of the protein sequence is done in FASTA format, or accession number entry is done to get information of superfamily. The third stage screening of gene was done based on identified superfamilies and their conserved region. These conserved gene sequences were used for the construction of a hybrid gene sequence.

11.3 *In Silico* Approaches for Functional and Structural Characterization of Gene and Target Protein of Microbes Responsible for Waste Degradation

11.3.1 Pairwise Sequence Alignment

Pairwise sequence alignment tools are used for two biological sequences of protein or nucleic acid. It is used to identify regions of similarity that may indicate functional, structural, and/or evolutionary relationships between two biological sequences. Tools such as EMBOSS Needle and SIM are used for pairwise sequence alignment. GeneWise tool is used to compare a protein sequence to a genomic DNA sequence.

11.3.2 Multiple Sequence Alignment

Multiple Sequence Alignment (MSA) is commonly the alignment of three or more biological sequences of protein or nucleic acid of almost similar length. The results of MSA can help to infer homology, the evolutionary relationships as a phylogenetic tree between the sequences studied. The tools such as Clustal W, Clustal Omega, COBALT, DbClustal, Kalign, EMBOSS Cons, MAFFT, MUSCLE, MView, T-Coffee, PROBCONS, webPRANK, GUIDANCE, SALIGN, AlignMe, and PRALINE are widely used for small, medium, and long MSA [23].

11.3.3 Evolutionary Relationship Analysis

The phylogenetic analysis was used to identify the novel microbes from the known microbes that are related to the degradation process, such as cyclic nitramine degradation [24], geosmin degradation [25], alkane degradation [26], and textile dye decolorization [27]. In polyethylene biodegradation, the degrading bacteria were identified by 16S rRNA sequencing. The identified bacterium for the polyethylene biodegradation was *Pseudomonas citronellolis*, and BLAST result with the database gave 96% similarity, which was the nonsignificant score; thus, it was clarified that the identified bacterium was a novel strain [28].

The phylogenetic analysis can be done using various *in silico* tools such as CVTree, PHYLIP, Phylogeny.fr, PHYML, Phylemon2, Phylodendron, Phylogenetic tree prediction, POWER, ProtTest, T-Rex, webPRANK, Replacement Matrix, and DIVEIN to determine the ancestors of the organisms [19]. The genome- and proteome-based global studies in environmental microbiology are opening a new era in accelerating and developing bioremediation technologies [18, 19, 29].

11.3.4 Structural Analysis

The two PDB protein structures are compared by using FATCAT pairwise alignment tool. This tool can also be used to search for similar structures for given PDB structure. Protein structures are flexible and undergo structural rearrangements as part of their function. It simultaneously addresses the two major goals of flexible structure alignment: optimizing the alignment and minimizing the number of rigid-body movements (twists) around hinges, which is introduced in the reference structure [30, 31].

11.4 *In Silico* Tools for Waste Degradation

11.4.1 Bioinformatics Tools for Waste-degrading Microbial Community Analysis

In silico bioinformatics tools help in identifying and characterizing microbial populations involved in waste degradation. Bioinformatics platforms help to study the genetic and functional properties of microbial communities used in waste degradation.

11.4.1.1 Metagenomic Analysis Platforms

Tools like QIIME2, MG-RAST, and MEGAN analyze high-throughput sequencing data from waste treatment systems through the process of next-generation sequencing (NGS) data to identify microbial taxa and functional genes in waste ecosystems such as anaerobic digesters or wastewater bioreactors. It helps to identify key degraders and predict metabolic potential for degrading xenobiotics, plastics, and heavy metals [32–34]. Biodrying of kitchen waste involves microbial degradation that reduces moisture and organic content. *In silico* tools like metagenomic analysis and KEGG pathway mapping were used to track microbial functions and enzyme activity changes during the 25-day process. These tools revealed key shifts in metabolic pathways and identified dominant microbes such as *Weissella, Aeribacillus*, and *Bacillus.* The findings support the role of *in silico* methods in optimizing biodrying through microbial process monitoring [35].

11.4.1.2 Genome Annotation and Comparative Genomics

Tools like RAST, Prokka, KEGG, and Pathway Tools are used to annotate genomes and infer functional pathways relevant to biodegradation by identifying enzymes involved in waste degradation (e.g., monooxygenases and dioxygenases). Identification of catabolic genes such as *alkB, pca*, and dioxygenases present in microbes is used for hydrocarbon or dye degradation. Researchers use metagenomic sequencing to profile microbial communities in hypersaline wastewaters.

11.4.2 Molecular Docking and Dynamics for Enzyme–Substrate Interactions

Molecular modeling simulates the interaction between pollutant molecules and enzymes or microbial proteins. There are various molecular docking tools that are available to study the potential binding interaction between waste and protein for biodegradation.

11.4.2.1 Molecular Docking Tools

The biodegradation capability of microbes on chemical can be identified by molecular docking study of protein–ligand. The binding site of the enzyme should be analyzed deeply; tools available are active site prediction, PAR-3D, Pocket-finder, PROCAT, and Q-SiteFinder. Freely available tools for docking studies are AutoDock, MOE, SwissDock, CB Dock, CB Dock 2, DOCK, GOLD, Glide, SCIGRESS, GlamDock, GEMDOCK, iGEMDOCK, HomDock, ICM, Flex-Ensemble (FlexE), Fleksy, FITTED, VLifeDock, ParaDockS, Molegro Virtual Docker, eHiTS, OEDocking, idock, Rosetta Ligand, rDock, Lead Finder, GalaxyDock, FINDSITE-LHM, ADAM, PLANTS, HADDOCK, Computer-Aided Drug Design, GriDock, DockoMatic, and BDT [36]. It is reported that AutoDock is comparatively better than DOCK, Flex, and GOLD, whereas Schrödinger is one of the best and most efficient tools among all for molecular docking analysis [37].

The molecular docking software is used to simulate and predict the binding of pollutants such as toxic compounds, plastics, dyes, pesticides, heavy metals, and various other wastes with the biodegradative enzymes such as lipases, laccase, hydrolase, peroxidase, and many more. It helps to predict efficiency of enzymes or engineered mutants for pollutant degradation. The biodegradation of the pollutants can be screened by using various available protein–ligand docking tools against the available protein, protein–DNA interaction, protein–protein interaction, and 3D docking study. Docking algorithms can easily predict the possible interactions between functional protein and ligand; it can also help to identify the involved hidden mechanisms. This information can help in producing the metabolically engineered organism that can enhance the degradation of wastes [38, 39].

The various *in silico* docking methods are used for studying degradation processes such as azo dye degradation [40] and pyrene degradation [41]. The pollutants are collected from the PubChem database, such as phenols and anilines, and docked with laccase enzymes to study effective bioremediation [19]. The research was carried out for biodegradation of textile azo dye Remazol Yellow RR using non-adapted bacteria *Lysinibacillus sphaericus* MTCC 9523. The molecular docking study was conducted for azoreductase and laccase of *L. sphaericus* against azo dye to correlate with decolorization and degradation percentage. The enzyme–dye active site analysis revealed a high dock score of −29.2622 and −21.2777 kJ/mol for azoreductase and laccase, respectively [42].

The computational studies are carried out to hypothesize the degradation possibility, probable mechanism, and the interaction between microbial lipase and polyethylene and polystyrene, which are the major polymeric substances present in plastics [43]. The 3D structures of lipase (PDB ID: 2LIP) [44] and extracellular lipase (PDB ID: 2Z8X) [45] were retrieved from PDB database for Pseudomonas sp.; similarly, the 3D structures of polystyrene and polyethylene were retrieved from PubChem database [46]. The molecular docking was carried out by the PatchDock server to study protein–ligand interactions [47], and the docked complexes were visualized using PyMOL [48].

The minimum binding energy and score for oxidized polyethylene with extracellular lipase were −166.94 kcal/J and 4,022, respectively. The main residues identified were Tyr 29, His 30, Asn 31, Leu 32, Gly 60, Thr 61, Gln 64, Gly 65, Ile 70, Arg 106, Gln 120, and Arg 141 for the ligand binding. The network of hydrogen bonding and covalent interactions shows stability to the interactions. Similarly, ligand polystyrene was also docked with the extracellular lipase and was showing minimum energy of −235.96 kcal/J and a score of 5,382. The main interacting residues found were Ser 28, Tyr 29, Leu 32, Asn 62, Ser 63, Gly 65, Asn 104, Arg 106, Thr 108, Gln 120, and Arg 147. Based on the interaction between microbial lipase and selected polymeric compounds via computation studies, it was suggested to further validate through experimental analysis [49].

The indigo dye was decolorized using mixed bacterial strains. The interaction study was carried out for ThrH protein (PDB ID: 1RKU) of *Pseudomonas aeruginosa* PA01 and indigo dye using molecular docking by Schrodinger software. The ligands range was accepted based on both glide energy (−26.4 Kcal/mol) and glide score (−4.84 Kcal/mol) by XP and SP docking methods. The satisfactory hydrogen bond (1.86 and 1.93 Å) interaction was also observed within the ligand's molecules. The docking studies gave the understanding for better performance regarding the indigo dye degradation. This is the most innovative, efficient, and applicable technique for degradation process [50].

11.4.2.2 Molecular Dynamics Simulations

Molecular dynamics (MD) tools and software like GROMACS, AMBER, or CHARMM simulate enzyme flexibility and protein stability under different environmental conditions such as pH and temperature. These tools are used to evaluate enzyme robustness in bioreactors for waste treatment. Researchers used docking and MD simulations to study interaction between textile dye and microbial enzyme laccase. The decolorization of commercial dyes was studied using recombinant laccases from *Yersinia enterocolitica* and *Bacillus pumilus*. The laccase genes were cloned, overexpressed in *Escherichia coli*, and purified from inclusion bodies. The enzymes showed alkali tolerance, thermostability, and efficient dye decolorization. Molecular docking and MD simulations confirmed strong interactions between laccases and dyes, which were experimentally validated, showing up to 90.4% decolorization efficiency [51].

11.4.3 Metabolic Pathway Reconstruction and Simulation

11.4.3.1 Pathway Prediction and Analysis

Genome-scale metabolic models (GSMMs) simulate cellular metabolism and predict waste degradation pathways. Several pathway prediction systems are also available along with the known pathway databases. These prediction pathway systems help in predicting pathways of those chemicals for which the degradation pathways are not known in the literature. The enzymes involved in degradation pathways can also be predicted by using these pathway's prediction systems [52]. A few metabolic pathway databases are Biocyc, BRENDA, BioCarta, KEGG, MetaCyc, MetaRouter, PANTHER, OptCom, Pathway Commons, Roche Biochemical Pathways, UM-BBD EMP, and WIT. The KEGG database is the highest-praised and most user-friendly database, though ExPASy, PathDB, and UM-BBD databases are also very useful [53]. Tools like KEGG, MetaCyc, and OptCom pathway tools also help to model degradation routes of complex compounds. These tools support genome-scale metabolic modeling (GSMM) of microbial consortia. Researchers used GSMM to model a microbial consortium degrading aromatic compounds [54, 55].

11.4.3.2 Flux Balance Analysis

Flux Balance Analysis (FBA) is a mathematical method for simulating the metabolism of cells or entire unicellular organisms, such as *E. coli* or yeast, using genome-scale reconstructions of metabolic networks. Software like COBRA Toolbox simulates metabolite flow under different nutrient or pollutant loads. These tools help to simulate metabolic fluxes in consortia degrading complex waste such as lignocellulose and hydrocarbons. These tools can be utilized to optimize nutrient supplementation in microbial consortia for dye or plastic degradation [54, 55].

MetQuest is a scalable graph-based algorithm that identifies all possible metabolic pathways, including branched and cyclic ones, in large genome-scale networks. It works efficiently even with minimal input and is applicable to microbial communities. Though it may overpredict due to network gaps, it is a valuable first-line tool for exploring metabolism and generating experimental hypotheses [56].

The research shows the exceptional metabolic adaptability of *Pseudomonas putida* in the biodegradation of industrial pollutants such as catechols, a phenolic compound. Utilizing computational tools, researchers identified 1,387 distinct metabolic routes, with up to 25 reaction steps, leading to fumarate, a central intermediate in the TCA cycle. This exhaustive mapping confirms the bacterium's ability to employ both ortho- and meta-cleavage pathways, validating its potential for biotechnological applications in environmental remediation. Such *in silico* approaches not only enhance our understanding of microbial catabolic capabilities but also reinforce role of *P. putida* as a promising agent for sustainable waste treatment technologies [57–60].

11.4.4 System Biology and Network Modeling

Systems biology uses computational and network modeling to offer powerful approaches to study the complex microbial interactions involved in waste biodegradation. By integrating omics data and pathway analysis, these methods help identify key enzymes, genes, and metabolic networks responsible for breaking down pollutants. Network modeling can reveal how microbial communities coordinate to degrade organic waste efficiently. This holistic understanding supports the development of optimized biodegradation strategies for effective waste treatment. Systems biology and network modeling also employ a variety of tools to visualize, analyze, and simulate complex biological systems.

11.4.4.1 Microbial Interactions

The tools include Cytoscape for network visualization, CellDesigner for diagram editing, and computational platforms such as SimBiology (MATLAB), Python, and COBRA for in-depth modeling and analysis. These tools help in visualizing and analyzing networks of microbial consortia to understand syntrophic relationships in anaerobic digesters or composting units. It helps to enhance understanding of bottlenecks in microbial activity and pollutant breakdown [61–63]. Few more tools, such as OmicsNet, VisANT, Pathway Studio, and Patika, are used for the microbial interactions.

11.4.4.2 Signal Transduction and Regulation

Signal transduction is essential for cellular communication and is central to many biomedical studies. Key challenges include identifying signaling molecules, understanding their context-specific interactions, and mapping where and when these events occur. Tools and databases like CellDesigner, KEGG, and BioModels help reconstruct signaling networks and analyze pathway dynamics. These resources offer structured insights but remain limited due to incomplete pathway knowledge [64]. Tools like CellDesigner simulate stress response or quorum sensing (QS) pathways triggered by pollutants. Researchers modeled regulatory networks in nitrifying bacteria in wastewater plants [65]. The activated sludge process plays a vital role in wastewater treatment, with sludge forming as flocs, biofilms, or granules. QS and quorum quenching (QQ) regulate microbial behavior and influence sludge structure through signaling molecules, extracellular polymeric substance (EPS), and microbial interactions. This study highlights the mechanisms of QS/QQ in shaping sludge forms and their roles in systems like biofilm reactors and membrane bioreactors. It also discusses current challenges and future directions for effective QS/QQ applications in wastewater treatment [66].

Systems biology and network modeling rely on a diverse range of tools to visualize, simulate, and analyze the complexity of biological systems. Among the most widely used visualization tools is Cytoscape, an open-source platform that enables the integration and graphical representation of molecular interaction networks. For more immersive and interactive 3D network exploration, OmicsNet provides a web-based solution, while other tools like VisANT, Pathway Studio, and Patika support the dynamic visualization of biological networks, incorporating elements such as gene expression data and time-series behavior. For model building and simulation, CellDesigner offers a structured interface for constructing biochemical and regulatory networks, using SBML (Systems Biology Markup Language) as a standardized format for sharing models. Computational platforms like MATLAB and Python play a significant role in analysis, with MATLAB offering specialized toolboxes for simulation and Python providing libraries such as NetworkX and Biopython for network and biological data analysis. Several simulation platforms like Virtual Cell (VCell), COPASI, BioNetGen, and NEURON are tailored for constructing and simulating detailed mathematical models of cellular and neural systems. Tools like EasyModel simplify the modeling process using ordinary differential equations, particularly for beginners, while PySCeS supports advanced metabolic modeling and control analysis. Petri nets serve as a graphical framework for analyzing reaction-based systems and are particularly suited for representing biological pathways. Supporting resources include the BioModels Database, a repository of curated SBML-format models; Reactome and KEGG, which provide detailed pathway databases; and NeAT, a toolkit for assessing structural properties of biological networks. Techniques such as flux variability analysis (FVA) and linear programming are employed to investigate and optimize metabolic fluxes. Metabolite tracing allows researchers to track biochemical transformations within metabolic pathways, while emerging standards like SPML and CellML bridge biological modeling with machine learning (ML) ecosystems, enhancing model interoperability and computational analysis. A strong understanding of biology, mathematics, and computational techniques, along with the use of advanced tools, is crucial for researchers engaged in systems biology and network modeling.

11.4.5 Artificial Intelligence and ML

Artificial Intelligence (AI) and ML play a vital role in improving waste biodegradation and management by enabling data-driven decision-making. AI/ML algorithms are increasingly integrated into waste treatment models and help in optimizing treatment processes, predicting outputs, and evaluating system performance. These technologies help predict waste composition, monitor microbial activity, and optimize treatment parameters in real time. ML models can also be used to design efficient waste segregation and recycling strategies. By automating analysis and forecasting, AI and ML enhance the overall efficiency and sustainability of waste management systems.

11.4.5.1 Process Optimization

Process optimization using AI and ML enables data-driven decision-making to improve efficiency, reduce costs, and enhance performance across various industries. AI/ML algorithms analyze large datasets to identify patterns, predict outcomes, and recommend optimal process parameters in real time. In fields like manufacturing, waste management, healthcare, and energy, these technologies help automate complex tasks, minimize resource usage, and adapt to changing conditions dynamically. Techniques such as predictive modeling, reinforcement learning, and neural networks are commonly applied for continuous process improvement and system optimization. ML models analyze spectral, image, or physicochemical data to classify and segregate solid waste. ML models such as ANN, SVM, Random Forest, and Decision Trees predict chemical oxygen demand/biological oxygen demand (COD/BOD) removal efficiency, nutrient removal, or biogas yield and optimize parameters like pH, temperature, and retention time. Researchers employed ANN to model anaerobic digestion (AD) of food waste and optimize biogas production. Toolkits such as TensorFlow, Scikit-learn, and MATLAB are used for various analyses [67].

11.4.5.2 Predictive Toxicology and Biodegradability

In silico tools help to predict the toxicity and biodegradability of various chemical substances. These computational tools and techniques are used to model, analyze, and predict the toxicological effects of chemical substances. These methods provide insights into how chemicals may interact with biological systems and potentially lead to harmful outcomes, aiding in risk assessment and regulatory decision-making. These methods offer efficient alternatives to traditional laboratory-based *in vitro* and *in vivo* experiments. By utilizing chemical databases, structure–activity relationship (SAR) models, and ML technologies, they support faster risk evaluation and compound screening. This approach not only streamlines chemical and various researches but also helps to minimize the sets of experiments.

Various *in silico* tools and databases are available to support toxicity prediction of chemicals through computational modeling. Tools like OECD, OPERA, VEGA, EPI Suite, ECOSAR, QSAR Toolbox and BIOWIN are commonly used to evaluate the environmental persistence, mobility, and toxicity of pharmaceutical and industrial compounds and are also used to model biodegradability and environmental fate of chemicals based on molecular descriptors. The *in silico* biodegradation prediction involves the use of computational approaches to estimate how effectively chemicals can be broken down by microorganisms in the environment. These methods rely on various models and tools, such as Quantitative Structure–Activity Relationship (QSAR) and Quantitative Structure–Biodegradation Relationship (QSBR) models, which predict biodegradability based on molecular structure. AI techniques, including neural networks and genetic algorithms, are also employed to simulate biodegradation pathways and environmental fate. Additionally, molecular docking and

MD simulations provide insights into enzyme–substrate interactions during microbial degradation processes. Expert systems and rule-based models apply established biochemical knowledge to assess the likelihood of degradation. These tools have broad applications, including environmental risk assessment by predicting chemical persistence, supporting bioremediation strategies through identification of suitable degrading microorganisms, and aiding in the development of environmentally friendly products by designing readily biodegradable compounds. They also assist in improving wastewater treatment by forecasting the performance of biological treatment systems. Some practical tools used in this field include Biodeg, which estimates ready biodegradability, and computational methods for assessing ligninolytic enzyme activity, contributing to our understanding of microbial degradation capabilities.

Solid waste management (SWM) involves multiple factors, including technical, environmental, financial, and regulatory aspects. Recently, AI has emerged as a promising approach to address SWM challenges through advanced computational methods. This review guides researchers on the use of AI in SWM by exploring various models, their advantages, limitations, and practical applications. It also compares AI approaches with traditional methods and highlights specific SWM areas where AI has been applied. The discussion concludes with current advancements, challenges, and future prospects of AI-driven SWM solutions [68].

11.4.6 Computational Fluid Dynamics

Computational Fluid Dynamics (CFD) plays a significant role in understanding and optimizing waste degradation processes. CFD tools simulate fluid flow, mass transfer, heat transfer, gas generation, and mixing within waste treatment systems like composting reactors, landfills, or biodigesters. CFD helps predict degradation efficiency. It enables detailed analysis of oxygen distribution, temperature profiles, and microbial activity zones, which are crucial for effective biodegradation. CFD models assist in designing and scaling up reactors for enhanced waste breakdown and reduced environmental impact. This approach minimizes the need for trial-and-error in real systems, saving time and resources. Overall, CFD supports sustainable and efficient waste management strategies. Kinetics and CFD modeling have been extensively applied to analyze biological treatment systems. A review of studies from 2006 to 2023 categorizes integration strategies into three types: (i) directly coupled CFD-kinetics models, which offer detailed insights but are computationally intensive; (ii) uncoupled models, which reduce computational demand but are less suitable for real-time analysis sub-cycling is suggested for design purposes; and (iii) indirect coupling through CFD-based compartment models, providing a balanced and efficient approach suitable for real-time applications [69].

11.4.6.1 Reactor Design

Optimize aeration, sedimentation, and mixing patterns in moving bed biofilm bioreactors (MBBRs), sequencing batch reactors (SBRs), and trickling filters. The tools, such as ANSYS Fluent and COMSOL Multiphysics, simulate flow dynamics in reactors. The CFD tools are used to design and scale up bioreactors, lagoons, or trickling filters for optimal waste treatment. These tools simulate oxygen transfer and mixing in aerobic sludge treatment plants. This tool helps to simulate floc behavior and clarifier performance for settling sludge. Researchers have studied CFD modeling of mixing patterns in an anaerobic digester using ANSYS Fluent [70].

AD is an effective method for treating diverse waste types while generating biogas as a clean energy source. The performance of AD relies heavily on reactor design, which influences key factors like mixing, heat transfer, and microbial kinetics. Traditionally, optimizing these reactors through lab experiments is time-consuming and costly. CFD offers a cost-effective solution by simulating fluid flow and heat transfer, aiding efficient reactor design. This review highlights AD fundamentals and CFD applications and explores existing research gaps in reactor optimization and process integration [71].

11.4.7 Geographic Information Systems and Remote Sensing

Geographic Information Systems (GISs) and Remote Sensing (RS) are powerful tools increasingly applied in waste treatment and degradation management. GIS enables spatial analysis and mapping of waste generation, disposal sites, and transportation routes, which helps optimize collection and treatment strategies. GIS integrates spatial data and also helps to monitor and plan waste management activities. RS, through satellite imagery and aerial data, facilitates monitoring of illegal dumping, landfill dynamics, and environmental impacts such as leachate dispersion and vegetation stress. The integration of GIS and RS allows for real-time monitoring, site suitability analysis, and tracking of degradation processes, thereby improving decision-making and promoting sustainable waste management systems [72].

11.4.7.1 Landfill Sitting and Monitoring

GISs support the identification of landfilling site selection, monitoring of leachate movement, and mapping pollution hotspots. It helps in identifying optimal locations for landfills or treatment plants using land use, soil, and hydrology layers. It is done with applying toolkits such as ArcGIS and QGIS. The study highlights the significance of using RS and GIS for improving SWM in rapidly growing urban areas. By identifying optimal waste disposal sites and transport routes, this approach can reduce environmental impacts and public inconvenience. Applying the model to Gudivada town provides practical insights to enhance current waste management strategies and support sustainable urban planning. ArcGIS 10.3 version was used for this project [73].

11.4.7.2 Pollution Mapping

Pollution mapping is a crucial technique in waste treatment and management, as it helps visualize and analyze the spread of pollutants in the environment. By identifying contamination sources, affected zones, and pollution intensity, this approach supports informed decision-making for waste handling and remediation. It enables authorities to prioritize areas for cleanup, monitor the effectiveness of treatment processes, and prevent further environmental degradation. Pollution mapping, when integrated with digital tools like GIS, enhances the accuracy and efficiency of waste management strategies, promoting a cleaner and healthier ecosystem.

Satellite-based NDVI or TIR imaging assesses the impact of leachate or effluent discharge. It is also used to evaluate the impact of landfills or wastewater discharge using remote satellite data. Few researchers used GIS to evaluate leachate risk zones in Beirut landfill regions [74].

11.5 Construction of Genetically Engineered Microorganisms for Waste Treatment

A genetically engineered microorganism (GEM) or microbial consortia is a collective effort of biologists, computer scientists, engineers, and ecologists. The novel synthetic consortia capabilities help in developing unique microorganisms and ecologically benefit consortia. These advanced technologies will support the development of computational systems biology, which will benefit society through recycling and effective anthropogenic waste management. Genetic engineering is having a promising future in various streams. The *in silico* approaches can be applied in the construction and analysis of consortia for designing of GEMs. The *in silico* approach can help in reducing the cost by reducing the number of hits and trials in the construction of GEMs during wet laboratory experiments. The microbial consortia engineering is widely accepted, as it has the potential to generate a highly novel and effective biological tool for degradation of waste compounds [10]. The degradation capability of *in silico* designed GEM can also be studied using *in silico* approaches [11, 12]. *In silico* metabolic engineering of microbes is applicable in various microbiology-related fields such as agriculture, medicine, and many industries [13]. There are many *in silico* tools available for gene and protein data mining, such as NCBI, GenBank, DDBJ, and UniProt. They help in understanding of metabolic pathways for biodegradation and bioremediation through KEGG, Branda, Gene Ontology, and GEO [14–16]. The *in silico* tool such as Blastp can help in identifying homologous proteins and conserved regions of proteins that are responsible for degradation of various waste components. The Vector NTI Tool helps in the construction of hybrid ORF of screened proteins of waste-degrading genes. This tool can also help in analysis of constructed hybrid ORF to check the potential of hybrid protein designed. The designed hybrid protein can be modeled using Swiss modeler [75] and Schrödinger software suite [76–78]. Vector NTI tool was used to verify hybrid sequence and to design a clone for hybrid ORF [79]. The conserved regions of the gene sequence of selected superfamilies were used for the construction of a hybrid gene sequence. The hybrid gene sequence of the selected conserved region was imported in vector NTI, and after entering the hybrid sequence, it was analyzed for the parameters. The analysis of parameters done for constructed gene sequence was open reading frame analysis, restriction enzyme analysis, translation analysis, and thermodynamic analysis. The constructed gene sequence was cloned through

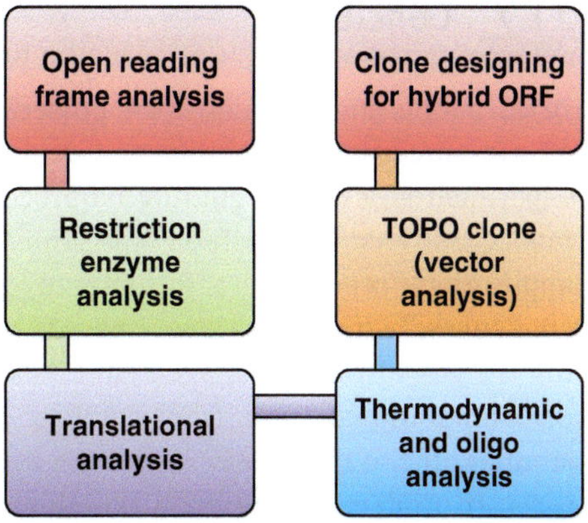

Figure 11.2 Flow diagram showing various analysis tools available for constructed hybrid ORF using vector NTI tool.

the various cloning tools, such as TOPO cloning tool. The sequences of analysis performed are shown in Figure 11.2.

11.6 Benefits of *In Silico* Approaches in Waste Treatment

- **Reduced Experimental Costs**
  - More cost-effective than traditional methods.
  - Minimizes lab-scale trial-and-error.
  - Minimizes the need for *in vivo* experiments.
- **Time Efficiency**
  - Faster to reach the results.
  - Rapid hypothesis testing and screening.
- **Predictive Insight**
  - Anticipates microbial behavior or pollutant fate.
- **Process Optimization**
  - Refines bioreactor parameters.
- **Environmental Impact Analysis**
  - Models' pollution dispersion and treatment success.
- **High-throughput Screening**
  - Enables rapid assessment of large compound libraries.
- **Mechanism-based insights**
  - Provides information on how chemicals interact with biological systems.
- **Emerging Trends**
  - **Digital Twins:** Virtual models of physical waste treatment plants, combining real-time data with simulations for predictive control.
  - **Cloud-based Platforms:** Integrated tools like KBase, Galaxy, and Google Earth Engine enable collaborative and scalable *in silico* waste modeling.
  - **Synthetic Biology Integration:** Design of synthetic microbial consortia or enzymes using *in silico* gene circuit simulation (e.g., GeneNet, Cello).

11.7 Conclusion

In silico tools and techniques are powerful enablers for modern waste treatment, offering precision, scalability, and deep system-level insight. They complement experimental work by guiding bioprocess design, enhancing microbial performance, and ensuring sustainability. Integrating these digital tools with field-scale applications represents a future-ready strategy for global waste management challenges. *In silico* tools have emerged as transformative instruments in addressing the global waste management crisis. This chapter highlighted a wide range of computational tools, from database mining and gene analysis to advanced simulation and modeling. Tools such as BLAST, KEGG, QIIME2, and AutoDock assist in understanding gene function, microbial interactions, and enzymatic efficiency. Predictive modeling using AI and ML enables real-time monitoring and process optimization in bioreactors and landfill systems. Molecular docking and dynamics simulations provide insights into pollutant–enzyme interactions, guiding enzyme engineering for higher degradation efficiency. CFD tools are crucial in simulating and improving reactor designs to enhance microbial activity and reduce environmental impacts. GIS and RS allow for effective spatial analysis, pollution mapping, and informed site selection. Through pathway reconstruction and metabolic modeling, researchers can uncover hidden degradation routes and simulate microbial behavior. *In silico* tools also assist in assessing the toxicity and biodegradability of industrial chemicals, enabling safer waste treatment strategies. These approaches reduce reliance on extensive laboratory testing, saving time and resources. They also support the development of synthetic biology solutions, like engineered microbial consortia. Combining computational tools with experimental validation fosters a data-driven and targeted approach. Integration with systems biology allows comprehensive analysis of microbial networks and signal regulation. Such insights help design robust and sustainable waste degradation systems tailored for specific pollutants. This interdisciplinary toolkit supports both academic research and industrial application in environmental biotechnology. It empowers scientists to explore novel degradation mechanisms and microbial capabilities with precision. As environmental challenges intensify, digital tools will play an increasingly central role in developing greener solutions. Future trends such as digital twins, cloud platforms, and AI-driven workflows will further enhance waste treatment efficiency. Integrating *in silico* predictions of both toxicity and biodegradability offers a more complete evaluation of a chemical's safety and its potential environmental effects. The *in silico* tools are also available for the construction of GEMs, such as Vector NTI Tool. This comprehensive strategy supports the identification of substances that are both less harmful to human health and more environmentally friendly. Ultimately, *in silico* techniques pave the way for a cleaner, healthier, and more sustainable planet.

References

1 Shivani, G. (2017). Bioremediation of agricultural, municipal, and industrial wastes. In: *Handbook of Research on Inventive Bioremediation Techniques* (ed. J.N. Bhakta), 341–363. IGI Global.
2 Reineke, W. and Knackmuss, H.J. (1988). Microbial degradation of haloaromatics. *Ann Rev Microbiol*. 42: 263–287.
3 Streit, B. (1992). Bioaccumulation processes in ecosystems. *Experientia*. 48 (10): 955–970.
4 Seeger, M., Hernández, M., Méndez, V., et al. (2010). Bacterial degradation and bioremediation of chlorinated herbicides and biphenyls. *J Soil Sci Plant Nutr*. 10 (3): 320–332.
5 Mrozik, A., Piotrowska-Seget, Z., and Labuzek, S. (2003). Bacterial degradation and bioremediation of polycyclic aromatic hydrocarbons. *Pol J Environ Stud*. 12 (1): 15–25.
6 Spain, J.C. (1995). Biodegradation of nitroaromatic compounds. *Ann Rev Microbiol*. 49: 523–555.
7 Venulet, J. and Van Etten, R.L. (1970). *The Chemistry of the Nitro and Nitroso Groups* (ed. H. Fever), 201–289. New York: Interscience.
8 Vargas, J.M. (1975). Pesticide degradation. *J Arboric*. 1 (12): 232–233.
9 Chawla, N., Suneja, S., Kukreja, K., and Kumar, R. (2013). Bioremediation: an emerging technology for remediation of pesticides. *Res J Chem Environ*. 17 (4): 88–105.
10 McMurry, J. (2000). Aromatic hydrocarbons. In: *Organic Chemistry*, 120–180. Brooks/Cole.
11 Raffi, F., Hall, J.D., and Cernigila, C.E. (1997). Mutagenicity of azo dyes used in foods, drugs and cosmetics before and after reduction by Clostridium species from the human intestinal tract. *Food Chem Toxicol*. 35: 897–901.
12 Vandevivere, P.C., Bianchi, R., and Verstraete, W. (1999). Treatment and reuse of wastewater from the textile wet-processing industry: review of emerging technologies. *J Chem Technol Biotechnol*. 72: 289–302.
13 Lloyd, J.R. and Lovley, D.R. (2001). Microbial detoxification of metals and radionuclides. *Curr Opin Biotechnol*. 12: 248–253.
14 Banerjee, A., Jhariya, M.K., Yadav, D.K., and Raj, A. (2018). Micro-remediation of metals: a new frontier in bioremediation. In: *Handbook of Environmental Materials Management* (ed. C. Hussain), 1–36. Cham: Springer.
15 Alexander, M. (1965). Biodegradation: problems of molecular recalcitrance and microbial fallibility. *Adv Appl Microbiol*. 7: 35–80.
16 Garg, M. Xenobiotic compounds: meaning, hazards and biodegradation. www.biologydiscussion.com/microbiology2/bioremediation/xenobiotic compounds-meaning-hazards-and-biodegradation/55625 (accessed 03 July 2019).
17 Leblanc, R. (2019). The decomposition of waste in landfills: a story of time and materials. The Balance Small Bussiness.

18 Watson, J.D., Baker, T.A., Bell, S.P., et al. (2004). *Molecular Biology of the Gene*, 293–342. Pearson Education Inc.
19 Umadevi, S., Aalfin, E.S., Ayyasamy, P.M., and Rajakumar, S. (2015). Computational approaches in waste management: special emphasis in microbial degradation. *Res Rev: J Ecol Environ Sci*. Special issue (Integrated Waste Management and Energy Recovery-S1): 22–27.
20 Obayori, U.S. (2009). Differential degradation of crude oil (Bonny Light) by four Pseudomonas strains. *J Environ Sci*. 21 (2): 243–248.
21 Copley, S.D. (2009). Evolution of efficient pathways for degradation of anthropogenic chemicals. *Nat Chem Biol*. 5: 559–566.
22 Mount, D.W. (2000). *Bioinformatics: Sequence and Genome Analysis*. New York: Cold Spring Harbor Laboratory Press.
23 Brennan, E. (2009). Bioinformatic analysis and in vitro site directed mutagenesis of conserved amino acids of $BphK^{LB400}$, a specific bacterial glutathione transferase. *Int Biodeter Biodegr*. 63: 928–932.
24 Zhao, J. (2004). Phylogeny of cyclic nitramine-degrading psychrophilic bacteria in marine sediment and their potential role in the natural attenuation of explosives. *FEMS Microbiol Ecol*. 49 (3): 349–357.
25 Hoefel, D. (2009). Biodegradation of geosmin by a novel Gram-negative bacterium; isolation, phylogenetic characterization and degradation rate determination. *Water Res*. 43 (11): 2927–2935.
26 Miralles, G. (2007). Alkane biodegradation and dynamics of phylogenetic subgroups of sulfate-reducing bacteria in an anoxic coastal marine sediment artificially contaminated with oil. *Chemosphere*. 68 (7): 1327–1334.
27 Palanivelan, R. (2014). Optimization of prime parameters for textile dye decolorization by design of experiments (DOEs) using *Lysinibacillus fusiformis* M1. *Desalin Water Treat*. 56 (4): 944583. https://doi.org/10.1080/19443994.2014.944583.
28 Bhatia, M. (2014). Implications of a novel *Pseudomonas* species on low density polyethylene biodegradation: an *in vitro* to *in silico* approach. *SpringerPlus*. 3: 497.
29 Bansal, A.K. (2005). Bioinformatics in microbial biotechnology – a mini review. *Microb Cell Fact*. 4 (19): 254.
30 Ye, Y. and Godzik, A. (2004). FATCAT: a web server for flexible structure comparison and structure similarity searching. *Nucl Acid Res*. 32: 582–585.
31 Li, Z., Ye, Y., and Godzik, A. (2006). Flexible structural neighborhood—a database of protein structural similarities and alignments. *Nucl Acid Res*. 34: 277–280.
32 Meyer, F., Paarmann, D., D'Souza, M., et al. (2008). The metagenomics RAST server–a public resource for the automatic phylogenetic and functional analysis of metagenomes. *BMC Bioinform*. 9: 1–8.
33 Bolyen, E., Rideout, J.R., Dillon, M.R., et al. (2019). Reproducible, interactive, scalable and extensible microbiome data science using QIIME 2. *Nat Biotechnol*. 37 (8): 852–857.
34 Vavourakis, C.D., Ghai, R., Rodriguez-Valera, F., et al. (2016). Metagenomic insights into the uncultured diversity and physiology of microbes in four hypersaline soda lake brines. *Front Microbiol*. 7: 211.

35 Cai, L., Guo, H.T., Zheng, G.D., et al. (2022). Metagenomic analysis reveals the microbial degradation mechanism during kitchen waste biodrying. *Chemosphere.* 307: 135862.

36 Park, H. (2006). Critical assessment of the automated AutoDock as a new docking tool for virtual screening. *Proteins.* 65: 549–554.

37 Morris, G.M., Huey, R., Lindstrom, W., et al. (2009). AutoDock4 and AutoDockTools4: automated docking with selective receptor flexibility. *J Computat Chem.* 30 (16): 2785–2791.

38 Bitetti-Putzer, R. (2001). Functional group placement in protein sites: a comparison of GRID and MCSS. *J Comp Aid Mol Des.* 15: 935–960.

39 Kroemer, R.T. (2007). Structure based drug designing docking and scoring. *Curr Protein Pept Sci.* 8: 312–328.

40 Sridhar, S. and Chandra, J.H. (2014). Involvement of computational tools towards *In Silico* remediation - synthetic textile dyes interacting with azoreductase. *Int J ChemTech Res.* 6: 4412–4416.

41 Arun, A. (2008). Polycyclic aromatic hydrocarbons (PAHs) biodegradation by Basidiomycetes fungi, Pseudomonas isolate, and their cocultures: comparative in vivo and in silico approach. *Appl Biochem Biotechnol.* 151 (2–3): 132–142.

42 Srinivasan, S. (2018). Biodegradation of textile azo dye, Remazol Yellow RR using non- autochthonous bacteria Lysinibacillus sphaericus MTCC 9523, supported by docking. *Proceedings of the International Conference on Biodiversity and Sustainable Resource Management.* 2018 Nov.

43 Tokiwa, Y., Calabia, B.P., Ugwu, C.U., and Aiba, S. (2009). Biodegradability of plastics. *Int J Mol Sci.* 10 (9): 3722–3742.

44 Schrag, J.D., Li, Y., Cygler, M., et al. (1997). The open conformation of a Pseudomonas lipase. *Structure.* 5 (2): 187–202.

45 Angkawidjaja, C., You, D.J., Matsumura, H., et al. (2007). Crystal structure of a family I.3 lipase from Pseudomonas sp. MIS38 in a closed conformation. *FEBS Lett.* 581 (26): 5060–5064.

46 Sussman, J.L., Lin, D., Jiang, J., et al. (1998). Protein Data Bank (PDB): database of the three-dimensional structural information of biological macromolecules. *Acta Crystallogr.* 54 (Pt 6 Pt 1): 1078–1084.

47 Schneidman-Duhovny, D., Inbar, Y., Nussinov, R., and Wolfson, H.J. (2005). Patch Dock and Symm Dock: servers for rigid and symmetric docking. *Nucl Acids Res.* 33: 363–367.

48 Seeliger, D. and de Groot, B.L. (2010). Ligand docking and binding analysis with PyMOL. *J Comp-Aid Mol Des.* 24 (5): 417–422.

49 Skariyachan, S., Rizvi, A., Kamath, M., and Vasist, K.S. (2015). Selection and screening of microbial consortia for efficient and ecofriendly degradation of plastic garbage collected from urban and rural areas of Bangalore, India. *Environ Monit Assess.* 187 (1): 4174.

50 Prabhavathi, P., Rajendran, R., Karthik Sundaram, S., et al. (2016). Molecular docking studies on potent adsorbed receptor of Thrh protein: a new target for biodegradation of indigo dye. *J Bioremediat Biodegrad.* 7: 356. https://doi.org/10.4172/2155-6199.1000356.

51 Ahlawat, S., Singh, D., Virdi, J.S., and Sharma, K.K. (2019). Molecular modeling and MD-simulation studies: fast and reliable tool to study the role of low-redox bacterial laccases in the decolorization of various commercial dyes. *Environ Pollut.* 253: 1056–1065.

52 Arora, K. and Bae, H. (2014). Integration of bioinformatics to biodegradation. *Biol Proced Online.* 16: 8.

53 Wittig, U. and Beuckelaer, A.D. (2001). Analysis and comparison of metabolic pathway databases. *Brief Bioinform.* 2: 126–142.

54 Karp, P.D., Paley, S.M., Krummenacker, M., et al. (2010). Pathway Tools version 13.0: integrated software for pathway/genome informatics and systems biology. *Brief Bioinform.* 11 (1): 40–79.

55 Zomorrodi, A.R. and Maranas, C.D. (2012). OptCom: a multi-level optimization framework for the metabolic modeling and analysis of microbial communities. *PLOS Comput Biol.* 8 (2): e1002363.

56 Ravikrishnan, A., Nasre, M., and Raman, K. (2018). Enumerating all possible biosynthetic pathways in metabolic networks. *Sci Rep.* 8 (1): 9932.

57 Kumar, A. (2005). Pseudomonas putida: a versatile biocatalyst for environmental clean-up. *Indian J Microbiol.* 45 (1): 7–10.

58 Feist, C.F. and Hegeman, G.D. (1969). Phenol and benzoate metabolism by *Pseudomonas putida*: regulation of tangential pathways. *J Bacteriol.* 100 (2): 869–877.

59 Kanehisa, M., Furumichi, M., Tanabe, M., et al. (2017). KEGG: new perspectives on genomes, pathways, diseases and drugs. *Nucl Acids Res.* 45 (D1): D353–D361. https://doi.org/10.1093/nar/gkw1092.

60 Karthikeyan, S., Prathap, M.S., Kirubanandan, S., and Mahesh, S. (2020). Insights into wastewater treatment using AI models. *Bioresour Technol.* 305: 123476. https://doi.org/10.1016/j.biortech.2020.123476.

61 Funahashi, A., Morohashi, M., Kitano, H., and Tanimura, N. (2003). CellDesigner: a process diagram editor for gene-regulatory and biochemical networks. *Biosilico.* 1 (5): 159–162.

62 Funahashi, A., Matsuoka, Y., Jouraku, A., et al. (2008). CellDesigner 3.5: a versatile modeling tool for biochemical networks. *Proc IEEE.* 96 (8): 1254–1265.

63 Shannon, P., Markiel, A., Ozier, O., et al. (2003). Cytoscape: a software environment for integrated models of biomolecular interaction networks. *Genome Res.* 13 (11): 2498–2504.

64 Lindvall, J.M., Emelie, K., Blomberg, M., and Smith, C.E. (2003). In silico tools for signal transduction research. *Brief Bioinform.* 4 (4): 315–324.

65 Espinosa, J., Mairet, F., Kim, M., et al. (2018). Toward a regulatory network of nitrogen metabolism in nitrifying bacteria. *NPJ Syst Biol Appl.* 4: 23.

66 Wang, X., Yi, K., Pang, H., et al. (2024). An overview of quorum sensing in shaping activated sludge forms: mechanisms, applications and challenges. *Sci Total Environ.* 927: 171886.

67 Karthikeyan, M., Vijayachitra, S., Pyingkodi, M., et al. (2022). Study on water quality and wastewater treatment using IOT. *2022 6th International Conference on Computing Methodologies and Communication (ICCMC).* 2022 Mar 29. 431–438. IEEE.

68 Vyas, S., Dhakar, K., Varjani, S., et al. (2023). Solid waste management techniques powered by in-silico approaches with a special focus on municipal solid waste management: research trends and challenges. *Sci Total Environ.* 891: 164344.

69 Sadino-Riquelme, M.C., Donoso-Bravo, A., Zorrilla, F., et al. (2023). Computational fluid dynamics (CFD) modeling applied to biological wastewater treatment systems: an overview of strategies for the kinetics integration. *Chem Eng J.* 466: 143180.

70 Kumar, A., Sharma, M.P., and Yadav, S. (2019). CFD modeling of mixing patterns in an anaerobic digester using ANSYS Fluent. *Bioresour Technol Rep.* 7: 100203.

71 Farid, M.U., Olbert, I.A., Bück, A., et al. (2025). CFD modelling and simulation of anaerobic digestion reactors for energy generation from organic wastes: a comprehensive review. *Heliyon.* 11 (2): e41911.

72 Kumar, A., Agrawal, A., and Rajawat, A.S. (2016). Use of remote sensing and GIS in solid waste management: a case study of Delhi. *Procedia Environ Sci.* 35: 209–217.

73 Haripavan, N. and Dey, S. (2023). Application of remote sensing and geographic information system in solid waste management for Gudivada Municipality, Andhra Pradesh, India. *Waste Manag Bull.* 1 (3): 128–140.

74 Al-Hajj, A., El-Fadel, M., and El-Hougeiri, N. (2017). Geospatial modeling for leachate risk assessment in landfill areas: a case study in Beirut. *Waste Manag.* 63: 256–267.

75 Swiss model. https://swissmodel.expasy.org (accessed 23 February 2019).

76 Schrödinger. (2014). *Prime. "Version 3.5."* New York, NY: LLC.

77 Andrei, A., Ivanov, D., Barak, A., and Jacobson, K. (2009). Evaluation of homology modeling of G protein-coupled receptors in light of the A2A adenosine receptor crystallographic structure. *J Med Chem.* 52: 3284–3292.

78 Ferrara, P. and Edgar, J. (2007). Evaluation of the utility of homology models in high throughput docking. *J Mol Model.* 13: 897–905.

79 Guoqing, L. and Moriyama, E.N. (2004). Vector NTI, a balanced all-in-one sequence analysis suite. *Brief Bioinform.* 5 (4): 378–388.

12

Cluster and Consortia Development for Treatment of Waste

12.1 Introduction

The advanced microbial processes, which help in the breakdown of undesirable substances, are termed as bioremediation. These techniques utilize the capability of microorganisms in reducing the toxicity and concentration of many xenobiotic pollutants [1, 2]. Some microorganisms have great ability to degrade these xenobiotics without any human augmentation, though sometimes external efforts are made to speed up the biodegradation process. If nutrients and oxygen are applied to the systems to improve their effectiveness and to accelerate biodegradation, it is known as the biostimulation process. Bioaugmentation is the adding of microorganisms in the systems, and if the biodegradation is complete, then it is referred to as mineralization [3].

The modulation of the genetics of these microorganisms can also speed up the biodegradation process. This has enabled analyses of environmentally relevant microorganisms, providing insights into biodegradation pathways and the ability of organisms to adapt to the changing environmental conditions. Therefore, research should focus on the understanding mechanism of the interaction between microorganisms and xenobiotic compounds in the environment that must include biochemical as well as genetic engineering areas. Such effective strategies can play a vital role in successful environmental cleanup by a diverse group of microorganisms for the efficient remediation of toxic xenobiotic compounds [2, 4]. These are the eco-friendly, economical, and efficient treatment techniques such as mixed population, cluster, and consortia system are developing quite rapidly in the field of environmental restoration.

12.1.1 Biodegradation Using a Mixed Population of Microorganisms

Generally, microbes involved in biodegradation of the xenobiotics are bacteria and fungi. The application of these microbes in the form of a mixed population is

Biological Processes and Techniques in Zero Waste Management, First Edition. Garima Awasthi and Mohit Nigam. Published 2026 by Wiley-VCH GmbH.

recommended for biodegradation. The use of a mixed population of microbes is found to be more efficient and faster due to the mechanism of attacking two different microbes on two different points through two different mechanisms in the same molecule. It also creates a condition of co metabolism [5].

12.1.2 Cluster and Consortia System

The clusters and consortia are such technologies that are developed for the treatment of waste. The collaborative approaches of microbes as clusters and consortia enhance efficiency and effectiveness during treatment of waste.

The microbial cluster is the group of closely related microorganisms that are clustered together based on the similar characteristics of similar members. These microorganisms are often identified through analysis techniques. The microbial clusters may not necessarily have strong interactions between members. Microbial clustering is often used as a data analysis technique to group similar microbes. Microbial clusters can be used to identify distinct microbial populations within a larger community. Microbial clusters can be used to identify distinct microbial populations within a larger community. Analyzing a gut microbiome sample and identifying a group of bacteria with similar genetic profiles, which could be considered a bacterial cluster [6].

The microbial consortium is a community of different microbial species that actively interact with each other, often displaying synergistic relationships and performing complex functions together within an environment. The microbial consortium is a functional community with interdependent members. Microbial consortium relies on cooperative interactions between different species for optimal function. Microbial consortium focuses on understanding the complex interactions and metabolic pathways within a diverse microbial community. The microbial consortium is often explored for their potential in bioremediation, wastewater treatment, and other applied fields where diverse functions are needed. The community of bacteria in the human gut includes different species working together to digest complex carbohydrates, with some bacteria break down large molecules into smaller ones, which can be further utilized by other species [7].

12.1.3 Clustered Systems

Cluster-based systems consolidate waste from multiple sources and process it at a central facility within a given locality. These systems are adaptable, allowing for modified designs to meet specific waste treatment requirements. They are particularly advantageous in rural or semi-urban areas where establishing a traditional centralized sewer system is not feasible. Clustered wastewater treatment setups, which handle effluent from several residences or buildings in regions where linking to a central sewer infrastructure is either impractical or too costly, are also known as decentralized systems. These systems typically use a network of narrow pipes to channel wastewater to a nearby treatment unit. After processing, the treated water is usually released into the environment through natural absorption into the soil. This method is cost-efficient and well-suited for communities outside dense urban zones. Moreover, beyond the physical grouping of infrastructure, biological clustering, such as deploying microbial consortia, adds another layer to effective waste treatment by enhancing biodegradation through diverse microbial activity [8–11].

12.1.4 Gene Cluster

Demerec and Hartman (1959) coined the term gene cluster; it is a group of closely related genes that code for the enzyme series that is the part of the metabolic pathway. The process for the formation of gene clusters is the duplication of genes [12]. The attributes of the microbes being used for biodegrading can be changed, modified, or enhanced by changing their gene structure and thus altering the enzyme patterns. Now, several genes share similar genetic features. If there are two or more genes within an organism's DNA that encode for similar polypeptides or proteins and collectively share a generalized function, such a group is called a gene cluster. Portions of the DNA sequence of the genes within a gene cluster may be found as identical. Therefore, a gene cluster may be used to assess the evolutionary relationship between organisms.

The salient features, which are attributed to gene clusters, are as follows [13]:

a. They are observable on the same chromosome, as they are found so close.
b. Their dispersion is random.
c. The distance between genes can vary.
d. DNA, which occurs between each repeat in the cluster, is non-conserved.
e. The homogenization of gene clusters is only possible by gene conversion.
f. Genetic complexity is not the outcome as they evolve over a long time.

The above identities are found to exist in all the genes of a gene cluster, which represents certain portions of a DNA sequence. The formation of gene clusters and their persistence can be explained with the help of various models. An algorithmic problem is found to exist when detecting gene clusters. Various approaches employed for *in silico* identification and prediction of genes include:

a. Finding Open Reading Frames (ORFs).
b. Homology search, which is mainly based on pairwise alignment.
c. Content-based methods, which are *ab initio* methods and mainly use statistics.
d. Signal-based methods dependent mainly on signals in the vicinity of coding regions.
e. Combination of these methods.

The gene cluster *tmoABCDE* in *Pseudomonas mendocina* KR1 was constructed for toluene-4-monooxygenase degradation. A generalized function is shared by the consortia of genes that are located within the DNA of an organism [14]. Researchers have also identified the pcaRKF gene cluster from *Pseudomonas putida* for the biodegradation and transportation of 4-hydroxybenzoate [15]. The pca genes are cloned, sequenced, and arranged in a single gene cluster *pcaIJFBDKCHG* from

Acinetobacter calcoaceticus, which is inducible by protocatechuate [16]. Researchers have constructed a new and unique gene cluster, which consists of *ORF18, ORF17, tesI, tesH, ORF11, ORF12*, and *tes DEFG* in *Comamonas testosteroni* TA441 for steroid degradation [17].

Thus, with data mining, the genes involved in biodegradation of a wide variety of toxic compounds were found in different microbial species. The clustering of genes with specificity for the type of compound as well as the category of microorganism can be done with the use of various interactive and analytical tools.

12.1.5 Microbial Consortia

Microbial consortia involve the use of diverse microorganisms that work synergistically to degrade complex waste materials. These consortia can be naturally occurring or synthetically assembled to target specific pollutants. For instance, combining microalgae and bacteria has been shown to enhance wastewater treatment efficiency and nutrient recovery. These consortia can effectively remove contaminants such as ammonia and phosphates from wastewater, contributing to improved water quality.

Microbial consortia play a key role in improving the speed and quality of composting in solid waste management. Research shows that these microbial combinations can often outperform traditional methods like vermicomposting in breaking down organic matter. When communities adopt decentralized treatment approaches using these microbial systems, they can create flexible and environmentally friendly waste management strategies suited to local requirements.

12.1.6 Mixed Microbial Consortia

Organic pollutants are metabolized by many bacteria, but the capability to degrade organic compounds is not possessed by a single bacterium. To degrade complex mixtures of toxic compounds, mixed microbial communities are most successful. This is due to the powerful biodegradative potential constituted by genetic information [5].

For the first time, a bacterial consortium of seven members were isolated from creosote-contaminated soil, which was capable of utilizing fluoranthene and could co-metabolically biotransform High Molecular Weight (HMW) polycyclic aromatic hydrocarbons (PAHs), which have more than four aromatic rings [18].

Based on available literature, it has been found that anaerobic conditions are more favorable for the decolorization of azo dyes by microbes. Biological treatments using a combination of aerobic and anaerobic methods in which microbial consortia are used to degrade dyes are reported in literature [19]. Mixed consortia of *Pseudomonas* sp., *Enterococcus* sp., and *Proteus* sp. have been used for decolorization and biodegradation of dyes [20]. The microbial consortium *Escherichia coli, Pseudomonas aeruginosa*, and *Acinetobacter* sp. from paper mill pulp and *Serratia* sp. (three isolates) and *Pseudomonas fluorescens* from the tannery are identified for the degradation, based on 16S rDNA sequence analysis [21]. Another microbial consortium of *Micrococcus diversus, Deinococcus radiophilus*, and *Alloiococcus otitis* was also developed for the degradation of pulp and paper mill effluent, and the microbial strains were continuously enriched under alkaline conditions in the presence of 4-chlorosalicylic acid present in effluent [22].

A consortium of *Mycobacterium, Pseudomonas,* and *Alcaligenes* sp. is reported for the mineralization of fluoranthene. Actinomycetes are mostly used as PAH-degrading bacteria. Few non-actinomycete bacteria are also reported to metabolize benz[a]anthracene, chrysene, fluoranthene, and pyrene [23]. The mixed culture of *P. putida, P. aeruginosa, Flavobacterium* sp., and an unidentified strain isolated from the soil are capable of metabolizing fluoranthene and pyrene as an organic carbon [24].

The chlorinated benzoic acid, dibenzofuran, phenoxyacetic acid, and salicylic acid are degraded by a potentially stable consortium of *Klebsiella pneumoniae, P. fluorescens, P. mendocina, and Pseudomonas cichorii.* These strains are also capable of degrading chlorinated aromatic compounds by ortho- and meta-ring cleavage [25]. A strategy for reducing environmental pollution by chlorinated dioxins and dibenzofurans was developed in the chemostat in which microbial populations in a sediment core contaminated by chlorinated aromatics were enriched, and a stable microbial consortium was obtained in the presence of 4-chlorosalicylic acid as sole carbon source [26].

The researchers selected three microbial strains, such as *Bacillus safensis, Bacillus licheniformis*, and *Fusarium oxysporum*, to apply as a consortium for higher rates of lignocellulose degradation after 10 days of treatment [27].

There are various studies that focus on the biodegradation of two forms of plastics, LDPE (low-density polyethylene) and HDPE (high-density polyethylene), by means of novel bacterial consortia screened from various sources [28].

12.2 Difference Between the Clusters and Consortia

The microbial clusters and consortia have some differences between them that are mentioned in Table 12.1.

Table 12.1 Difference between cluster and consortia.

S. No.	Clusters	Consortia
1.	Cluster is a group of closely related microorganisms	Consortium is a group of different microorganisms
2.	Grouping based on similar members	Grouping based on interdependent members
3.	Actively interact due to similar characteristics	Actively interact with each other, and together within an environment
4.	No need to figure out synergistic relationships to perform functions due to microbial similarity	Need to figure out synergistic relationships and perform complex functions

(Continued)

Table 12.1 (Continued)

S. No.	Clusters	Consortia
5.	May not have strong interactions between members	Always have cooperative interactions between different species for optimal function
6.	Identify different microbes within a larger community	Explored potential microbes for bioremediation, wastewater treatment, and other applied fields
7.	May or may not require diverse functions	Required for diverse functions
8.	Group of bacteria with similar genetic profiles	Group of bacteria with different genetic profiles
9.	Identify group of bacteria of similar species suppose from gut sample	Identify bacteria of different species suppose from gut sample

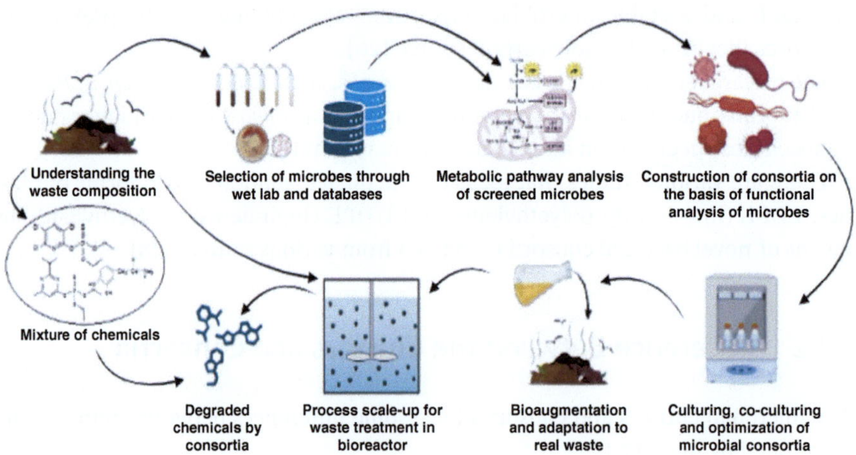

Figure 12.1 Process for developing microbial consortia and waste degradation.

12.3 Process for Developing Microbial Consortia

The process for developing microbial consortium for waste degradation involves various steps. The various steps of process involve understanding nature of waste, selecting microbial strains, and designing and optimizing a group of microorganisms that work together to break down complex waste materials efficiently (Figure 12.1).

The various key steps involved during the process of consortia development are as follows:

1. **Understanding the Waste Composition**
   The waste can be degraded only when the composition of waste is well understood. It is also required to know the complex chemical structure of components present in waste material, so that the chemical can be broken down with

the help of protein secreted from selected microorganisms used for construction of consortia. To understand the composition and characteristics of various chemicals present in waste material, the following points should be considered during the analysis:

- Analyze the physical and chemical properties of the waste.
- Analyze chemical reactions and pathways involved in breakdown of chemicals present in waste material.
- Identify key organic and inorganic components present in waste material.
- Determine factors like pH, moisture, and temperature.
- Determine potential toxins present in waste material.

Waste composition refers to the analysis of various materials that constitute solid waste generated from different sources such as households, industries, agriculture, and commercial establishments. Understanding the composition of waste is crucial for developing effective waste management strategies, enhancing resource recovery, and minimizing environmental impact.

Scientifically, waste can be broadly categorized into biodegradable (organic) and nonbiodegradable (inorganic) components. The biodegradable fraction typically includes food waste, yard trimmings, paper, and agricultural residues. These materials undergo microbial decomposition and are suitable for composting or anaerobic digestion, leading to biogas and compost production [29]. The nonbiodegradable fraction includes plastics, metals, glass, and synthetic materials, many of which are recyclable or require specialized disposal methods.

Compositional Studies Often Involve Physicochemical Characterization, Including:

- **Moisture Content:** Affects the energy potential and decomposition rate.
- **Calorific Value:** Determines the energy yield during incineration or pyrolysis [30].
- **Elemental Analysis:** Identifies carbon, hydrogen, nitrogen, sulfur, and oxygen content, which is vital for thermochemical conversion processes.
- **Heavy Metal Content:** Critical for assessing environmental risks and leachate toxicity, especially in landfill scenarios [31].

Variations in waste composition are influenced by geography, economic status, season, and local practices. Urban areas tend to produce more packaging materials and e-waste, while rural areas generate more organic agricultural residues.

Understanding waste composition is foundational for designing integrated solid waste management and treatment strategies, optimizing recycling processes, biodegradation process through clustering and consortia development, and fermentation process for achieving zero waste and sustainability goals [31].

2. **Selection of Microbial Strains**

   The screening of microbes for the construction of clusters and consortia can be done through *in silico* or *in vitro* studies. In *in silico* methods various databases are used for the screening of microbes on the basis of the responsible gene, enzyme, and the metabolic pathways for the degradation of particular waste. In *in vitro* methods the different microbes are screened from the identified source, and finally pure strains are selected using plating methods. The selection of microbes is done on the basis of metabolic compatibility of the strains.

- Identify microbes (bacteria, fungi) with specific enzymatic capabilities for degrading different waste components.
- Select strains based on their metabolic compatibility and synergy.
- Common microbes used for the construction of consortia are as follows:
  - **Bacteria:** *Pseudomonas, Bacillus, Acinetobacter* (hydrocarbon degradation) [32, 33]
  - **Fungi:** *Aspergillus, Trichoderma* (lignocellulose breakdown) [34]

The researchers have explored direct and standardized protocol for screening and selection of efficient e-waste-utilizing bacteria *in situ* to develop the potential consortia even for the degradation of e-waste. The biodegradation of e-waste by the selected strains during enrichment and *in situ* experiment was confirmed by FTIR, TG-DTG-DTA, and SEM analysis [35]. The various potential microorganisms are screened and identified as per the metabolic compatibility through *in silico* and *in vitro* methodologies, and further, they are selected for the construction of consortia for a particular process such as biodegradation, fermentation, or production formation.

3. **Construction of Consortium on the Basis of Functional Analysis**
The consortium is constructed on the basis of functional analysis. The microbes are combined as per the analysis of metabolic pathways and also the association and interaction between the microbes that are benefitting each other. The consortium can be constructed with the help of *in silico* and *in vitro* analysis.

- Combine microbes with complementary metabolic pathways (e.g., hydrolytic, fermentative, and methanogenic groups in anaerobic digestion).
- Ensure species interaction leads to mutual benefits (syntrophy), preventing competition.

Natural Consortium Construction Process

The researchers have developed a new way to build natural functional microbial consortia that can break down pollutants directly in the environment on the basis of metabolic pathway analysis. The phenanthrene was used as a test compound; they carefully selected and grew helpful bacteria like *Achromobacter* and *Pseudomonas*. These microbes worked better in real soil conditions than traditionally chosen ones. The findings show that microbes picked from the environment, not just the lab, are more reliable for cleaning pollutants. This method could help design better systems for sustainable waste cleanup and environmental health. This study can be extended to investigate functional capabilities of archaea and fungi also, to construct consortia for recycling and degrading pollutants in natural environment. These microbes, such as bacterial cells, fungi, and archaea, can be precisely identified, isolated, and grown individually, directly from their natural environment and cultured in *in situ* condition or optimized media condition. This approach helps to better understand microbial function with the help of gene sequencing and applying *in silico* tool. This *in vitro* and *in silico* approach helps to open new opportunities to explore useful resources [36].

Artificial Consortium Construction Process

Consortia are developed to form cyanobacterial–bacterial communities capable of effectively cleaning oil-contaminated soils and water. Pure cultures of *Phormidium* sp. K1 and *Oscillatoria* strains were isolated, and 14 associated bacterial strains were found to degrade hydrocarbons. These bacteria, mainly from the genera *Pseudomonas, Alcaligenes, Arthrobacter*, and *Bacillus*, were either isolated from natural sources or obtained from collections. Artificial consortia formed using these microbes showed strong remediation potential. Field tests confirmed that combinations like *Pseudomonas* K1 with *Pseudomonas stutzeri, Pseudomonas* sp. N2, and *Pseudomonas alcaligenes* significantly enhanced oil degradation, suggesting a promising microbiological approach for environmental cleanup [37].

4. **Microbial Consortia Culturing and Optimization in Laboratory**

   Microbial consortia are developed and maintained under laboratory conditions to enhance their functional efficiency. Culturing and co-culturing these consortia also aid in assessing their interactions and overall stability. Optimization aims to enhance degradation efficiency through various parameter modulations during the culturing of microbes and microbial consortia. A microbial consortium includes multiple symbiotic microorganisms that can endure varied environments by forming structures like biofilms [38–40].

   - The selected strains are grown under controlled conditions to test interactions and stability.
   - Optimize the following growth parameters:
     - **pH and Temperature:** Growth conditions are optimized for temperature, pH, and aeration before formulating the bioinoculant. It is required to adjust the pH and temperature provided to consortium up to tolerance ranges (e.g., thermophilic consortia for compost).
     - **Aeration and Agitation:** It is needed to optimize and ensure sufficient oxygen for aerobic consortia. Even the optimization is requirement of oxygen level (aerobic vs. anaerobic) for consortia culturing [41, 42].
     - **Substrate Concentration:** The substrate concentration is gradually increased to avoid toxicity shock. Consortium preparation involves culturing selected strains in nutrient or LB broth at $28 \pm 2\,°C$ with agitation for 24 hours to obtain high biomass [43].
     - **Carbon/Nitrogen Ratio:** It is required to optimize and balance nutrient availability to enhance microbial activity of consortia during culturing and degradation of waste.
     - **Biostimulation:** The additions of surfactants or micronutrients are also optimized to accelerate degradation process through microbial consortia.
     - **Statistical Tools:** Response Surface Methodology (RSM) and Design of Experiments (DoE) are commonly used for process optimization [44, 45].
     - **Co-culturing Techniques:** These are optimized to enhance degradation efficiency.

5. **Bioaugmentation and Adaptation**

   Bioaugmentation involves introducing specialized microbial clusters or consortia into contaminated environments to enhance biodegradation. Microbial

clusters are closely associated species or strains that coexist and often exhibit **syntrophic or synergistic interactions,** while **consortia** are broader, structured microbial communities designed or selected for specific biodegradation pathways [46, 47]. **Bioaugmentation** using consortia is more effective than single-strain inoculation for the following reasons:

- **Metabolic Complementarity:** Different strains degrade intermediates or co-metabolize substrates (e.g., hydrocarbons, xenobiotics).
- **Robustness:** Clusters offer resilience to environmental stress via microbial cooperation and cross-feeding.
- **Functional Redundancy:** Ensures degradation activity even if some microbes are lost due to stress or competition.
- **Adapt Microbial Consortia to Real Waste Conditions:** Gradual exposure to real waste conditions allows microbial strains to adapt metabolically and physiologically, improving survival and degradation efficiency. Adaptive evolution techniques, such as serial subculturing in selective environments, can enhance tolerance and enzymatic pathways for recalcitrant pollutant degradation.
  - **Sequential Enrichment Cultures:** This process is done by using the target pollutant as the sole carbon source to promote specialized cluster formation.
  - **Coevolutionary Incubation:** This process is under real effluent or solid waste matrices to develop stable microbial interactions.
  - **Horizontal Gene Transfer (HGT):** This process within clusters can enhance the spread of catabolic plasmids.

A synthetic consortium of *Pseudomonas, Bacillus*, and *Acinetobacter* sp. adapted to diesel-contaminated soil showed enhanced degradation due to cooperative alkane monooxygenase expression [48].

- **Use Genetic Engineering or Adaptive Evolution Techniques:**
  Genetic engineering is also used to insert specific catabolic genes into microbes for enhancing and improving the ability of microbes to degrade persistent organic pollutants (POPs). Recombinant *P. putida* expressing plasmid-borne degradation pathways has been effective against aromatic hydrocarbons [49].
- **Test Efficiency Through Small-scale Trials:**
  Small-scale trials are essential to assess community stability, degradation efficiency, and synergistic interactions before large-scale application.

6. **Field Application and Scale-up**

   Upon adaptation, microbial consortia are transferred to **field-scale systems.** The laboratory validation is done to scale up microbial consortia for field applications in systems like **bioreactors, composting units, or wastewater treatment plants.** The success of scale-up depends on maintaining *functional clustering*, i.e., microbial groupings that exhibit complementary metabolic activities and maintain functional stability and performance consistency of the consortium under variable environmental conditions.

   - The consortia are introduced into bioreactors, composting systems, or wastewater treatment plants. In **bioreactors,** biofilms often form naturally organized clusters with oxygen and nutrient gradients that influence microbial spatial structuring and enzymatic efficiency. **Co-metabolic clusters** in wastewater systems degrade complex organics (e.g., phenols, PAHs) through sequential enzymatic steps distributed among different species. A study using a mixed microbial consortium in an SBR (sequencing batch reactor) for azo dye removal reported that metabolic clusters involving *Sphingomonas* and *Comamonas* significantly improved degradation and process stability [50].

- Monitor performance using biochemical and molecular techniques (e.g., metagenomics, enzyme assays). The **metagenomic analyses** are done to assess shifts in microbial community structure and gene function. Metagenomic and transcriptomic tools track community structure, ensuring core degradative clusters persist over time. **Enzyme assays** are performed to quantify functional degradation enzymes such as laccases, peroxidases, dehydrogenase, and oxidoreductase to assess microbial metabolism. The interspecies communication is maintained that regulates enzyme expression and cluster cohesion [51]. Finally, the performance monitoring is conducted using physicochemical analysis of effluent or compost parameters such as Chemical Oxygen Demand (COD), Biological Oxygen Demand (BOD), and Total Suspended Solids (TSS).
- Scale-up process parameters are optimized based on degradation rates and microbial stability. **Optimization** involves adjusting pH, temperature, aeration, and nutrient supplementation to maintain degradation kinetics and microbial balance.
- The scale-up hurdles are reduced by using **carrier materials** such as biochar and alginate beads to promote colonization and immobilization of key clusters [52, 53].

7. **Monitoring and Sustaining the Consortium**

   To ensure long-term efficiency, microbial consortia must be routinely monitored for both structural stability and functional performance [54]. Long-term application requires **cluster-level monitoring** to assess:

   - **Microbial Interactions**: Identify species and mutualists maintaining biodegradative function [55].
   - **Functional Gene Distribution**: Use of quantitative PCR (qPCR) and metatranscriptomics to monitor expression of degradative operons (e.g., *nah, alkB*, and *cat*).

   The monitoring techniques include [54]:

   - Regularly analyze microbial diversity and activity using:
     - **PCR and Sequencing for Microbial Composition:** High-throughput sequencing is done to map microbial networks such as **PCR and 16S rRNA sequencing** for taxonomic profiling and detecting shifts in community dynamics [56].
     - **Gas chromatography** or **HPLC** is used for identifying intermediate and end products of biodegradation.
     - **Enzyme assays and gas chromatography–mass spectrometry (GC-MS)** for direct assessment of biodegradation end products and intermediary metabolites.
     - **Fluorescence *in situ* hybridization (FISH)** to spatially resolve microbial clusters.

- Maintain favorable conditions and prevent the dominance of undesired microbes: Environmental conditions (pH, nutrients, temperature) must be managed to prevent overgrowth of opportunistic or nonfunctional strains that can reduce degradation efficiency. Periodic supplementation or selective pressure may help maintain desired microbial populations.

Cluster Management Strategies:

- **Cluster Stability:** Assessed through co-occurrence network analysis to detect shifts or disruptions in core microbial modules.
  - Reinoculation of key functional clusters if loss is detected.
  - Avoidance of toxic shocks that can disrupt cluster integrity.

In a biofilter degrading volatile organic compounds (VOCs), dominant clusters of *Rhodococcus, Mycobacterium*, and *Nocardia* were sustained via nutrient control, ensuring stable performance [57].

8. **Applications of Microbial Consortia in Waste Degradation**
   - **Plastic Biodegradation:** *Ideonella sakaiensis* 201-F6 is a bacterium capable of degrading polyethylene terephthalate (PET), a common plastic used in bottles and containers. It produces two enzymes PETase and MHETase that sequentially break PET into its monomers (2-hydroxyethyl) terephthalic acid and ethylene glycol [58].
   - **Oil Spill Cleanup:** These hydrocarbon clastic bacteria are pivotal in breaking down oil in marine environments. *Alcanivorax borkumensis* dominates in oil-contaminated seawater due to its ability to utilize alkanes. *Pseudomonas* sp. degrade a broader range of hydrocarbons, including aromatic compounds. *A. borkumensis* has genes encoding alkane monooxygenases. In bioremediation trials, these bacteria proliferate in oil-contaminated waters when nutrients like nitrogen and phosphorus are supplemented [59, 60].
   - **Agricultural Waste Degradation:** Microorganisms such as *Trichoderma reesei, Aspergillus niger*, and *Bacillus subtilis* produce cellulases, xylanases, and ligninases to break down complex lignocellulosic biomass into fermentable sugars. Enzyme cocktails from fungi like *T. reesei* have high cellulolytic activity, making them essential in the pretreatment of agricultural residues such as straw and corn stover [61].
   - **Wastewater Treatment:** Microbial consortia in activated sludge systems rely on complex microbial communities, including *Nitrosomonas, Nitrobacter, Zoogloea*, and filamentous bacteria, for the removal of organic matter, ammonia, and nitrates from wastewater. The researchers show that community structure affects treatment efficiency. Biofilm-forming species enhance sludge flocculation and nutrient cycling [62].
   - **Biogas Production:** Anaerobic consortia digestion is carried out in four stages: hydrolysis, acidogenesis, acetogenesis, and methanogenesis for methane production. Methanogens like *Methanosarcina* and *Methanosaeta* play a vital role in converting acetate and hydrogen into methane. The balance among microbial guilds (hydrolytic, fermentative, acetogenic, and methanogenic) determines methane yield. Co-digestion of manure with agricultural residues improves biogas output [63].

12.4 Conclusion

Microbial technologies offer a powerful, sustainable alternative to conventional waste treatment methods, addressing the growing burden of environmental pollution. By harnessing the synergistic actions of mixed microbial populations, clusters, and consortia, complex and persistent waste materials can be effectively degraded. These biologically driven systems not only mimic natural ecological processes but also allow for optimization through scientific innovation, such as metagenomics, bioaugmentation, and synthetic biology. The integration of these advanced techniques enhances the efficiency, specificity, and scalability of microbial waste treatment. As demonstrated by recent research and practical applications, microbial consortia represent a promising frontier in environmental biotechnology. Their deployment can significantly reduce the ecological footprint of waste, support resource recovery, and align with the principles of a circular bioeconomy. Ultimately, the development of robust, engineered microbial systems is essential for building sustainable waste management strategies and ensuring long-term environmental health.

References

1 Leitão, A.L. (2009). Potential of *Penicillium* species in the bioremediation field. *Int J Environ Res Public Health*. 6: 1393–1417.

2 Joutey, N.T., Bahafid, W., Sayel, H., and Ghachtouli, N.E. (2013). Biodegradation: involved microorganisms and genetically engineered microorganisms. In: *Biodegradation: Life of Science* (ed. R. Chamy), 289–320. IntechOpen.

3 Bennet, J.W., Wunch, K.G., and Faison, B.D. (2002). Use of fungi biodegradation. In: *Manual of Environmental Microbiology*, 2e (ed. C.J. Hurst), 960–971. ASM Press.

4 Awasthi, G., Kumari, A., Pant, A.B., and Srivastava, P. (2017). *In silico* identification and construction of microbial gene clusters associated with biodegradation of undesired toxic materials. *Microb Pathog*. 114: 340–343.

5 Fritsche, W. and Hofrichter, M. (2005). Aerobic degradation of recalcitrant organic compounds by microorganisms. In: *Environmental Biotechnology: Concepts and Applications* (ed. H.-J. Jördening and J. Winter). Weinheim, FRG: Wiley-VCH Verlag GmbH & Co. KGaA.

6 Lozupone, C.A., Stombaugh, J.I., Gordon, J.I., et al. (2012). Diversity, stability and resilience of the human gut microbiota. *Nature*. 489 (7415): 220–230.

7 Zhou, J., Deng, Y., Luo, F., et al. (2010). Functional molecular ecological networks. *MBio*. 1 (4): 10–128.

8 U.S. Environmental Protection Agency (EPA). *Decentralized Wastewater Treatment Systems: A Program Strategy*. https://www.epa.gov

9 National Environmental Services Center (NESC). *Cluster Systems for Small Communities*. https://www.nesc.wvu.edu

10 Tchobanoglus, G., Burton, F., and Stensel, H.D. (2003). Wastewater engineering: treatment and reuse. *J Am Water Works Assoc*. 95 (5): 201.

11 Use of Microbial Consortia in Waste Treatment. https://www.sciencedirect.com/topics/biochemistry-genetics-and-molecular-biology/microbial-consortium.

12 Demerec, M. and Hartman, P.E. (1959). Complex loci in microorganisms. *Annu Rev Microbiol.* 13: 377–406.

13 Pilar Francino, M. (2012). The ecology of bacterial genes and the survival of the new. *Int J Evol Biol.* 14: 394026.

14 Yen, K.M., Karl, M.R., Blatt, L.M. et al. (1991). Cloning and characterization of a *Pseudomonas mendocina* KR1 gene cluster encoding toluene-4-monooxygenase. *J Bacteriol.* 173 (17): 5315–5327.

15 Harwood, C.S., Nichols, N.N., Kim, M.K., et al. (1994). Identification of the pcaRKF gene cluster from *Pseudomonas putida*: involvement in chemotaxis, biodegradation, and transport of 4-hydroxybenzoate. *J Bacteriol.* 176 (21): 6479–6488.

16 Kowalchuk, G.A., Hartnett, G.B., Benson, A., et al. (1994). Contrasting patterns of evolutionary divergence within the *Acinetobacter calcoaceticus* PCA operon. *Gene.* 146: 23–30.

17 Horinouchi, M., Hayashi, T., Yamamoto, T., and Kudo, T. (2003). A new bacterial steroid degradation gene cluster in *Comamonas testosteroni* TA441 which consists of aromatic-compound degradation genes for seco-steroids and 3-ketosteroid dehydrogenase genes. *Appl Environ Microbiol.* 69 (8): 4421–4430.

18 Mueller, J.G., Chapman, P.J., and Pritchard, P.H. (1989). Action of a fluoranthene-utilizing bacterial community on polycyclic aromatic hydrocarbon components of creosote. *Appl Environ Microbiol.* 55: 3085–3090.

19 Lodato, A., Alfieri, F., Olivieri, G., et al. (2007). Azo-dye conversion by means of *Pseudomonas* sp. OX1. *Enzyme Microb Technol.* 41: 646–652.

20 Chaube, P., Indurkar, H., and Moghe, S. (2010). Biodegradation and decolorisation of dye by mix consortia of bacteria and study of toxicity on *Phaseolus mungo* and *Triticum aestivum. Asiat J Biotechnol Resour.* 1: 45–56.

21 Shah, S. and Thakur, I.S. (2002). Enrichment and characterization of a microbial community from tannery effluent for degradation of pentachlorophenol. *World J Microbiol Biotechnol.* 18 (7): 693–698.

22 Jain, R.K., Kapur, M., Labana, S., et al. (2005). Microbial diversity: application of microorganisms for the biodegradation of xenobiotics. *Curr Sci.* 89 (1): 101–112.

23 Boldrin, B., Tiehm, A., and Fritzsche, C. (1993). Degradation of phenanthrene, fluorene, fluoranthene and pyrene by a *Mycobacterium* sp. *Appl Environ Microbiol.* 59 (6): 1927–1930.

24 Trzesicka-Mlynarz, D. and Ward, O.P. (1995). Degradation of polycyclic aromatic hydrocarbons (PAHs) by a mixed culture and its component pure cultures obtained from PAH-contaminated soil. *Can J Microbiol.* 41 (6): 470–476.

25 Thakur, I.S. (2001). Biodegradation of chlorinated benzoic acid by a *Pseudomonas fluorescens* from a chemostat. *Indian J Microbiol.* 41: 27–32.

26 Thakur, I.S. (1995). Structural and functional characterization of a stable, 4-chlorosalicylic acid degrading, bacterial community in a chemostat. *World J Microbiol Biotechnol.* 11 (6): 643–645.

27 Kumar, S., Dhar, H., Nair, V.V., Bhattacharyya, J.K. et al. (2016). Characterization of municipal solid waste in high-altitude sub-tropical regions. *Environ Technol.* 37 (20): 2627–2637.

28 Skariyachan, S., Manjunath, M., Shankar, A., et al. (2018). Application of novel microbial consortia for environmental site remediation and hazardous waste management toward low- and high-density polyethylene and prioritizing the cost-effective, eco-friendly, and sustainable biotechnological intervention. In: *Handbook of Environmental Materials Management* (ed. C. Hussain), 431–478. Cham: Springer.

29 Kumar, S., Bhattacharyya, J.K., Vaidya, A.N., et al. (2017). Characterization of municipal solid waste in high-altitude sub-tropical regions of India: a case study of Sikkim. *Waste Manag.* 67: 132–143.

30 Singh, R.P., Tyagi, V.V., Allen, T., et al. (2020). An overview for exploring the possibilities of energy generation from municipal solid waste (MSW) in Indian scenario. *Renew Sustain Energy Rev.* 15 (9): 4797–4808.

31 Zhang, D., Keat, T.S., and Gersberg, R.M. (2019). A comparison of municipal solid waste management in Berlin and Singapore. *Waste Manag.* 29 (7): 912–921.

32 Diez, M.C. (2010). Biological aspects involved in the degradation of organic pollutants. *J Soil Sci Plant Nutr.* 10 (3): 244–267.

33 Seo, J.S., Keum, Y.S., and Li, Q.X. (2009). Bacterial degradation of aromatic compounds. *Int J Environ Res Public Health.* 6: 278–309.

34 Wu, Z., Peng, K., Zhang, Y., et al. (2022). Lignocellulose dissociation with biological pretreatment towards the biochemical platform: a review. *Mater Today Bio.* 16: 100445.

35 Debbarma, P., Zaidi, M.G., Kumar, S., et al. (2018). Selection of potential bacterial strains to develop bacterial consortia for the remediation of e-waste and its in situ implications. *Waste Manag.* 79: 526–536.

36 Li, J., Luo, C., Cai, X., et al. (2024). Microbial consortium assembly and functional analysis via isotope labeling and single-cell manipulation of polycyclic aromatic hydrocarbon degraders. *ISME J.* 18 (1): wrae115.

37 Zhubanova, A.A., Ernazarova, A.K., Kaiyrmanova, G.K., et al. (2013). Construction of cyanobacterial-bacterial consortium on the basis of axenic cyanobacterial cultures and heterotrophic bacteria cultures for bioremediation of oil-contaminated soils and water ponds. *Russ J Plant Physiol.* 60: 555–562.

38 Singh, R., Paul, D., and Jain, R.K. (2019). Biofilms: implications in bioremediation. *Trends Microbiol.* 17 (9): 389–397.

39 Shahi, A., Aydin, S., Ince, B., and Ince, O. (2016). Assessment of microbial community dynamics in bioreactors for treatment of complex wastewaters. *Sci Rep.* 6: 34355.

40 Zhang, S., Merino, N., Okamoto, A., and Gedalanga, P. (2018). Interkingdom microbial consortia mechanisms to guide biotechnological applications. *Microbial Biotechnol.* 11 (5): 833–847.

41 Zhang, J., Zeng, G., Chen, Y., et al. (2020). Optimization of microbial consortia for bioremediation using microbial community engineering. *Environ Sci Technol.* 54 (14): 8580–8590.

42 Ghosh, A., Dey, N., Bera, A., et al. (2018). Optimization of physicochemical parameters for crude oil degradation by microbial consortia. *Biocatal Agric Biotechnol.* 14: 54–62.

43 Gautam, S.S., Gondwal, M., Soni, R. and Gautam, B.P.S. (2022). Potash biofertilizers: current development, formulation, and applications. In: Trends of Applied Microbiology for *Sustainable Economy* (ed. R. Soni, D.C. Suyal, A.N. Yadav, R. Goel), 481–500. Elsevier.

44 Kumar, S., Sharma, J., Joshi, S., et al. (2022). Design of Experiments (DoE) for optimization of biodegradation parameters in industrial wastewater treatment. *J Environ Chem Eng.* 10 (5): 108487.

45 Fang, H., Liang, D., Zhang, T., et al. (2021). High-throughput evaluation of microbial consortia dynamics in response to environmental stress. *Microbiome.* 9 (1): 52.

46 Tyagi, M., da Fonseca, M.M.R., and de Carvalho, C.C.C.R. (2011). Bioaugmentation and biostimulation strategies to improve the effectiveness of bioremediation processes. *Biodegradation.* 22 (2): 231–241. https://doi.org/10.1007/s10532-010-9394-4.

47 Brenner, K., You, L., and Arnold, F.H. (2008). Engineering microbial consortia: a new frontier in synthetic biology. *Trends Biotechnol.* 26 (9): 483–489.

48 Zhou, H., Guo, W., He, Y., et al. (2019). Enrichment and characterization of microbial consortia from long-term hydrocarbon-contaminated soils for diesel biodegradation. *Sci Total Environ.* 686: 1200–1211.

49 Cases, I. and de Lorenzo, V. (2005). Genetically modified organisms for the environment: stories of success and failure and what we have learned from them. *Int Microbiol.* 8 (3): 213–222.

50 Kumar, A., Kumar, S., Kumar, A., and Singh, B. (2018). Synergistic interaction of microbial communities enhances biodegradation of azo dyes: a systems biology perspective. *J Hazard Mater.* 347: 248–258.

51 El-Kamah, H., El-Khateeb, M., and Elsayed, O. (2020). Enhancing microbial consortia stability and reusability by immobilization on natural carriers. *Environ Technol Innov.* 20: 101157.

52 Patel, K., Shah, M., and Gandhi, R. (2021). Pilot-scale bioreactor studies for treating dye and textile effluents using microbial consortia. *Environ Sci Pollut Res.* 28 (19): 23600–23613.

53 Zhang, F., Cheng, Z., Gao, B., et al. (2020). Application of microbial consortia for wastewater treatment: a review. *Chemosphere.* 246: 125708.

54 Liu, W., Wang, Y., Zhou, H., et al. (2020). Monitoring microbial community composition and functional genes during bioremediation of petroleum-contaminated soils. *Sci Total Environ.* 712: 136393.

55 Faust, K. and Raes, J. (2012). Microbial interactions: from networks to models. *Nat Rev Microbiol.* 10 (8): 538–550.

56 Wang, L., Zhang, J., Zhao, R., et al. (2017). Quantitative monitoring of microbial consortia for pollutant degradation using real-time PCR. *J Hazard Mater.* 321: 602–610.

57 Li, D., Li, Z., Bai, X., and Wu, D. (2020). Microbial community and functional clusters in a biofilter for toluene removal. *Bioresour Technol.* 298: 122562.

58 Yoshida, S., Hiraga, K., Takehana, T., et al. (2016). A bacterium that degrades and assimilates poly(ethylene terephthalate). *Science.* 351 (6278): 1196–1199. https://doi.org/10.1126/science.aad6359.

59 Varjani, S.J., Upasani, V.N., and Pandey, A. (2017). Bioremediation of petroleum oil contaminated soil and water. *Environ. Technol. Innov.* 6: 94–108.

60 Yakimov, M.M., Timmis, K.N., and Golyshin, P.N. (2007). Obligate oil-degrading marine bacteria. *Curr Opin Biotechnol.* 18 (3): 257–266. https://doi.org/10.1016/j.copbio.2007.04.006.

61 Lynd, L.R., Weimer, P.J., van Zyl, W.H., and Pretorius, I.S. (2002). Microbial cellulose utilization: fundamentals and biotechnology. *Microbiol Mol Biol Rev.* 66 (3): 506–577. https://doi.org/10.1128/MMBR.66.3.506-577.2002.

62 Seviour, R. and Nielsen, P.H. (2010). *Microbial Ecology of Activated Sludge.* IWA Publishing.

63 Angelidaki, I. and Sanders, W. (2004). Assessment of the anaerobic biodegradability of macropollutants. *Rev Environ Sci Bio/Technol.* 3: 117–129. https://doi.org/10.1007/s11157-004-2502-3.

13

Genetically Engineered Microorganisms for the Treatment of Waste

13.1 Introduction

Waste accumulation from industrial, municipal, and agricultural sources poses severe threats to ecosystems and human health. Traditional methods for waste disposal, such as landfilling, incineration, and chemical treatment, often lead to secondary pollution or are energy-intensive. Genetically Engineered Microorganisms (GEMs) have emerged as a powerful and sustainable tool for waste degradation and resource recovery. Through advanced genetic modifications, these organisms offer enhanced biodegradation capabilities and resilience in harsh environmental conditions [1]. GEMs offer a powerful and efficient approach for waste treatment, as they are intentionally modified to enhance the degradation of a wide variety of pollutants. By altering their genetic makeup, these microorganisms acquire the capability to break down an expanded range of waste materials, including hazardous and toxic compounds, with significantly improved speed and effectiveness compared to naturally occurring microbial strains. GEMs demonstrate superior degradation potential across numerous waste types and have proven valuable for bioremediation applications involving contaminants from diverse environmental sources. These engineered organisms possess a higher metabolic capacity for degrading pollutants and have shown successful results in controlled environments for the treatment of various harmful substances. The application of genetically modified microbes also offers a safer and more economically viable solution than many conventional waste treatment methods. Through the insertion of targeted genes encoding useful proteins, bacterial, fungal, or algal cells can be enhanced to improve their pollutant-degrading abilities. GEMs have been successfully used in the biodegradation of multiple environmental contaminants, such as oil spills, halobenzoates, naphthalenes, toluenes, trichloroethylene (TCE), octanes, and xylenes, showcasing their versatility and effectiveness in environmental cleanup efforts. Unlike their naturally occurring counterparts, these engineered strains exhibit superior degradation potential due to their enhanced adaptability and capacity for co-metabolism.

Biological Processes and Techniques in Zero Waste Management, First Edition. Garima Awasthi and Mohit Nigam. Published 2026 by Wiley-VCH GmbH.

By enabling quicker and more efficient breakdown of pollutants, genetic engineering supports sustainable environmental practices and contributes to improved public health outcomes [2].

13.2 Genetic Approaches for Biodegradation

Biodegradation, which occurs naturally, is much slower than the rates that are observed under ideal conditions in the laboratory. This is due to the fact that multiple factors collectively influence the rate at which natural biodegradation occurs. In order to effectively break down the specific contaminant, the introduced microorganisms must demonstrate a higher efficiency than the native microbial populations [3]. These supplemented microorganisms should possess the ability to rapidly adapt to the environmental conditions and exhibit high efficiency in degrading the pollutants within a defined or limited time frame [4]. This situation created the need for practical and effective technologies, which subsequently brought significant attention to the use of GEMs in bioremediation. These engineered microbes have shown successful results in breaking down a wide range of pollutants under specific and controlled conditions. Furthermore, GEMs possess an enhanced capacity for degradation, making them more efficient compared to naturally occurring microbial strains in targeted environmental applications [5].

13.3 Concept of GEMs

GEMs are microorganisms whose genomes have been modified to improve their functional abilities, especially for targeted waste degradation. This genetic manipulation may involve inserting novel genes, overexpressing specific enzymes, or deleting undesired metabolic pathways. Tools such as recombinant DNA technology and CRISPR-Cas9 have enabled precise gene editing, allowing the creation of microbes that can degrade complex or toxic compounds more efficiently [6]. These modifications can also enable the synthesis of useful by-products like biofuels or enzymes during the degradation process [7].

13.4 Genetically Modified Microorganisms

Genetically modified microorganisms (GMMs) are presently being utilized across a wide range of sectors, including agriculture, environmental bioremediation, healthcare, and various industrial applications [8]. GMMs possess the ability to break down organic compounds due to extensive research conducted on various degradation pathways, enzymatic reactions, associated enzymes, and the genes responsible for these processes. These comprehensive studies have facilitated the development of GMMs equipped with novel metabolic pathways. GMMs have been explored as an alternative to natural microbial strains that either degrade contaminants at a slow rate or are incapable of degrading certain pollutants. This advanced strategy has become essential for reducing the concentration of toxic substances in the environment and ensuring the preservation of environmental quality and ecological balance [8, 9].

Genetic modification can occur due to the mutation, either induced naturally or artificially. The mutation is a very slow process, still a potent source of genetic variation. Mutation can be induced into microorganisms by modifying their genetic structure; as a result, the active site of their enzymes shows an affinity to waste components. It has been reported that few microbes after mutation develop the ability to degrade waste components if they are continuously exposed to waste. New enzymatic pathways are also developed for xenobiotic degradation due to the mutations in the microorganism. The high level of accuracy has been observed during the breakdown of waste due to the modification of certain genes of microbes [10].

13.5 Genetically Engineered Microorganisms

GEMs have demonstrated significant potential in the bioremediation of contaminated soil, groundwater, and activated sludge, showcasing improved capabilities in degrading a broad spectrum of chemical pollutants. These microorganisms are developed through advanced genetic engineering methods that involve the modification of their genetic material, drawing inspiration from the natural genetic exchange mechanisms observed among microbial species. The methodologies used to create GEMs are commonly referred to as recombinant DNA technology, which allows for the introduction of specific genes that enhance the microorganisms' ability to break down various environmental contaminants [11]. Molecular biology helps in the construction of new pathways using catabolic segments of various microbes to increase the biodegradation capacity of microbes [12].

Genetic engineering is an advanced technological approach that creates synthetic microorganisms specifically designed to degrade particular contaminants. This process involves forming novel combinations of genes that do not naturally occur together in any organism. The most commonly employed techniques in this field include the insertion of single genes or operons, the construction of new metabolic pathways, and the modification or alteration of existing gene sequences to enhance the organism's degradation abilities [8, 13, 14].

Genetic engineering can be applied on a large scale for the application of the variety of microbial consortia available for biodegradation. The modulation of the genetics of these microorganisms can speed up the biodegradation process [15].

There are a minimum of four primary strategies utilized in the development of GEMs for use in bioremediation applications [16]. These strategies encompass: (i) Alteration of enzyme specificity and substrate affinity to enhance degradation efficiency; (ii) Construction and regulation of metabolic pathways to optimize pollutant breakdown; (iii) Development, monitoring, and control of bioprocesses to ensure effective and stable microbial activity; and (iv) Application of bioaffinity bioreporter sensors for the detection of chemicals, reduction of toxicity, and analysis of treatment endpoints.

13.5.1 Construction Process of GEMs

The creation of GEMs involves a structured sequence of steps that are specifically designed to introduce new characteristics or improve existing traits within microbial systems. This process leverages molecular biology tools to alter the genetic architecture of microorganisms, enabling them to perform targeted functions such as pollutant degradation, metabolite production, or biosensing. The construction of GEMs follows various steps. The initial step involves identifying and selecting an appropriate gene or group of genes relevant to the desired trait. In the second step, the chosen DNA fragment intended for cloning is inserted into a suitable vector. The third step involves introducing this vector into the host microbial cells, resulting in genetically modified bacteria referred to as recombinant cells. Following this, the process includes the amplification of the gene copies and the selection of host cells that successfully contain the recombinant DNA. In the final step, a screening procedure is carried out to identify clones that possess both the desired genetic material and the required biological characteristics [17–19]. The following is a detailed outline of the core stages involved in the construction of GEMs (Figure 13.1).

13.5.1.1 Gene Identification and Isolation

The first stage entails the identification of a gene that governs a particular trait of interest, such as the ability to degrade pollutants, produce specific enzymes, or resist heavy metals. After the gene has been located, it is extracted from its original organism through the use of molecular cloning methods to enable further genetic manipulation and integration into a new microbial host. Advanced bioinformatics

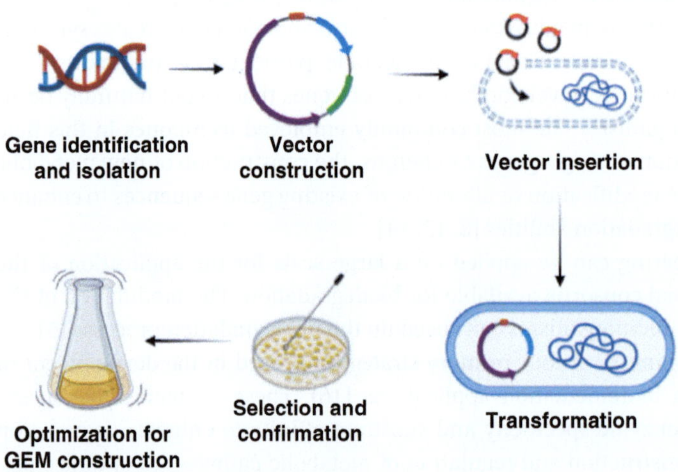

Figure 13.1 Steps involved in construction of GEMs.

tools can aid in the prediction and validation of gene function based on sequence homology and structural domains [20].

13.5.1.2 Vector Design and Construction

The extracted gene is subsequently introduced into an appropriate cloning vector, most commonly a plasmid, which acts as a vehicle for transferring the genetic material. To facilitate accurate and efficient expression of the gene within the host organism, crucial regulatory elements such as promoters, ribosome-binding sites, and transcription terminators are incorporated alongside the gene within the vector. Selection markers, such as antibiotic resistance genes, are incorporated to facilitate downstream identification of successful transformation [21].

13.5.1.3 Gene Transfer into Host Cells (Transformation)

The recombinant vector is transferred into a host microorganism (commonly *Escherichia coli, Pseudomonas*, or yeast species) through a process known as transformation. Various transformation techniques can be employed depending on the host, including:

- **Electroporation:** Involves the application of a controlled electric field to the cell membrane, temporarily creating pores that enable the uptake of foreign DNA into the cell.
- **Chemical Transformation:** Use of calcium chloride and heat shock to increase membrane permeability.
- **Natural Competence:** Exploiting the inherent ability of some microbes to uptake extracellular DNA [22].

13.5.1.4 Selection and Molecular Confirmation

After the transformation process, only the cells that have effectively taken up the recombinant plasmid are isolated, usually through the use of antibiotic selection markers. To verify the successful incorporation and structural accuracy of the inserted gene, various molecular techniques such as polymerase chain reaction (PCR), restriction enzyme digestion, and DNA sequencing are utilized. Expression analysis may also be conducted using SDS-PAGE, RT-PCR, or reporter assays [23].

13.5.1.5 Strain Optimization and Characterization

The final step involves optimizing the engineered strain to improve its performance. This may include:

- Enhancing gene expression through codon optimization or promoter engineering.
- Eliminating metabolic bottlenecks by modifying competing pathways.
- Adapting environmental or fermentation conditions to maximize yield and functionality.

Iterative cycles of genetic engineering, combined with systems biology and synthetic biology tools, can be used to fine-tune strain performance for industrial or environmental applications [24].

13.5.2 Plasmids and Vectors Used in the Construction of GEMs

The development of GEMs relies heavily on the use of plasmids and a variety of other vectors. These elements act as carriers for introducing foreign DNA into host cells, enabling targeted manipulation, replication, and expression of genetic material for research, industrial, or therapeutic purposes [25]. Bacteria, particularly species of *Pseudomonas*, serve as the primary focus for genetic manipulation efforts. These microorganisms are widespread in the environment and possess the natural ability to break down a wide range of toxic compounds. The genetic information responsible for the metabolism of these substances is found on both their chromosomes and plasmids. *Pseudomonas* species and similar bacteria represent a vital source of catabolic genes that are essential for applications in genetic engineering and the development of enhanced biodegradation capabilities [8, 26]. In genetic engineering, plasmids are frequently employed as cloning vectors to replicate or express specific genes. A vector is a genetic molecule that carries a collection of different genes, including those that confer antibiotic resistance. Vectors function by delivering new genetic material into recipient cells, where they can replicate independently of the host cell's chromosomal DNA. Additionally, transposons, another type of genetic element, can also serve as vectors, facilitating the transfer and integration of genes within or between genomes [26, 27].

13.5.2.1 Plasmids as Genetic Engineering Tools

Plasmids are small, circular, double-stranded DNA molecules that exist independently from the chromosomal DNA within bacterial cells. Due to their ease of manipulation and autonomous replication, they are widely used as cloning and expression vectors [20]. Nowadays, in the construction of GEMs, the artificial plasmid vectors are commonly used, these plasmids typically contain key elements such as:

- **Origin of Replication (oriC):** Facilitates independent replication within the host.
- **Selectable Markers:** Commonly antibiotic resistance genes for selecting transformed cells.
- **Multiple Cloning Site (MCS):** A region that includes a variety of restriction enzyme recognition sites, allowing for the convenient insertion of foreign genes into the vector.
- **Regulatory Sequences:** Including promoters, terminators, and enhancers to regulate gene expression.

These synthetic plasmids are engineered to incorporate the most advantageous features taken from various natural plasmids. Among them, expression plasmids are commonly utilized in modern genetic engineering practices. They enable the rapid and large-scale production of target proteins. In addition to plasmid vectors, enzymes play an essential and indispensable role as key tools in genetic engineering, particularly in cut-and-paste techniques. This group of enzymes includes restriction endonucleases, which cleave DNA at specific recognition sites, and DNA ligases, which are responsible for sealing breaks in the phosphodiester backbone of DNA strands. The commonly used restriction endonucleases in molecular biology are *Bam*HI, *Eco*RI and *Hind*III and T4 DNA ligase [13, 14, 17, 28]. Classic examples include **pBR322**, **pACYC177**, and **pACYC184**, which are engineered to carry MCS and resistance markers, enabling efficient gene insertion and selection [29]. TOL plasmids encode the enzymes for toluene degradation.

Recent innovations in vector design include modular plasmids compatible with CRISPR-Cas9, tunable promoters, and synthetic riboswitches, enhancing gene expression precision in microbial systems [30, 31].

13.5.2.2 Other Vector Types for GEM Construction

Depending on the application, alternative vectors may be more suitable, particularly when dealing with larger DNA fragments or host-specific requirements.

a. **Bacteriophages**
   Vectors like lambda phage are used in bacterial systems for cloning relatively large DNA fragments (~20 kb) and constructing genomic libraries [32].
b. **Cosmids**
   Combining plasmid and bacteriophage features, cosmids accommodate up to 45 kb of DNA, making them ideal for genome mapping and functional genomics [22].
c. **Artificial Chromosomes**
   - Bacterial Artificial Chromosomes (BACs) and Yeast Artificial Chromosomes (YACs) are utilized for the cloning of large DNA fragments, with BACs capable of carrying inserts larger than 300 kb and YACs accommodating DNA segments exceeding 1 Mb in size.
   - These are indispensable in genomic studies, transgenic organism development, and disease gene mapping [33].
d. **Integrative Vectors**
   Used in fungi, yeast, and some bacteria, integrative vectors insert genes directly into the host genome for stable expression. They are increasingly used in metabolic engineering and synthetic biology (Table 13.1) [34].

Table 13.1 Components of genetic vectors.

S. No.	Component	Function
1	Origin of replication	Allows independent replication in host cells
2	Multiple cloning site	Provides insertion points for foreign genes using restriction enzymes
3	Selectable marker	Enables identification of successful transformants (e.g., antibiotic resistance)
4	Promoter region	Initiates transcription of inserted genes
5	Regulatory sequences	Controls transcription and translation of cloned genes

13.5.2.3 Modern Vector Systems

Modern vector systems have evolved significantly beyond simple plasmid constructs to become highly sophisticated tools tailored for diverse molecular biology applications. These advanced vectors incorporate several innovative features to enhance their efficiency, versatility, and specificity in genetic manipulation across various organisms. Following are the key features:

a. **Fusion Tags for Protein Purification and Localization**
   Fusion tags are short amino acid sequences that are genetically linked to the protein of interest to aid in its purification, identification, or tracking within the cellular environment. These tags are commonly used to simplify downstream processing and to enhance the functional study of proteins. Typical examples include:
   - **His-tag (Polyhistidine tag):** A string of histidine residues (usually 6–10) that binds strongly to nickel or cobalt ions, enabling purification using immobilized metal affinity chromatography (IMAC).
   - **GFP (Green Fluorescent Protein):** Serves as a fluorescent tag that allows the visualization of protein expression and enables the tracking of subcellular localization in living cells.

b. **FLAG, HA, and Myc Tags**
   Epitope tags are recognized by specific antibodies for western blotting, immunoprecipitation, or immunofluorescence studies. Fusion tags can be attached to either the N-terminal or C-terminal end of the target protein and are frequently designed with specific protease cleavage sites, allowing their removal after purification to recover the native form of the protein.

c. **Shuttle Vectors**
   Shuttle vectors are specially designed genetic constructs capable of replicating and functioning in two or more distinct host species, usually including a prokaryotic host such as *E. coli* and a eukaryotic host like yeast, insect, or mammalian cells. These versatile vectors typically include:
   - Multiple origins of replication (one for each host)
   - Dual selectable markers for selection in both hosts
   - Promoters recognized by the host transcriptional machinery. Such versatility allows for easy cloning in bacteria followed by expression or functional analysis in more complex eukaryotic systems, streamlining workflows in synthetic biology and protein production.

d. **CRISPR-Compatible Systems for Precision Genome Editing**
   Modern vectors are now frequently integrated with CRISPR/Cas9 or related genome editing technologies. These systems are designed to facilitate:
   - Delivery of Cas9 or other endonucleases (e.g., Cas12a) along with guide RNAs (gRNAs) into cells.
   - Inducible expression of CRISPR components for controlled editing.
   - Integration of donor DNA templates designed for homology-directed repair (HDR), allowing for the precise introduction of specific mutations or the insertion of genes at targeted locations.

- Multiplexing capabilities to target multiple genomic sites simultaneously. These vectors often contain regulatory elements like U6 or T7 promoters for gRNA expression and CMV or EF1α promoters for Cas9 expression in mammalian systems. Additionally, newer generations of CRISPR vectors also incorporate base editors and prime editors for precise and programmable DNA alterations without double-strand breaks [31].

13.5.2.4 Vector Selection and Application

The choice of an appropriate vector is a critical step in genetic engineering and molecular biology experiments. Vectors serve as carriers for foreign DNA, and their selection directly influences the success of downstream applications such as gene expression, cloning, genome editing, and therapeutic delivery. The vector selected is based on following criteria:

a. **Size of the DNA Fragment**
   The capacity of a vector to accommodate DNA inserts varies depending on its type:
   - **Plasmid vectors** are ideal for small-to-moderate DNA fragments (up to ~10 kb) and are widely used in routine cloning and expression.
   - **Cosmids** and **BACs** can carry larger fragments, typically up to 45 kb and 300 kb, respectively, making them suitable for genome mapping or large gene cluster manipulation.
   - **YACs** are used when very large inserts (over 1 Mb) need to be maintained, such as in genomic library construction.

   Choosing a vector with the appropriate insert capacity ensures genetic stability and efficient replication or expression.
b. **Target Host Organism**
   Vectors must be compatible with the cellular machinery of the host organism where the DNA is to be replicated or expressed:
   - For **bacterial hosts** like *E. coli*, plasmids with bacterial origins of replication (e.g., ColE1) and selectable markers (e.g., ampicillin resistance) are commonly used.
   - For **yeast systems** such as *Saccharomyces cerevisiae*, shuttle vectors that contain both bacterial and yeast replication elements (e.g., 2μ origin, URA3 marker) are chosen.
   - For **non-model or industrial microbes** like *Pseudomonas* species, specialized broad-host-range plasmids (e.g., RSF1010 or RK2-derived vectors) are preferred for stable maintenance and gene expression.

   The vector must be optimized for replication, selection, and transcriptional activity in the chosen host to ensure experimental success.
c. **Purpose of Application**
   The intended application significantly influences the vector design and selection:
   - **Cloning Vectors** are designed primarily for the insertion and amplification of DNA fragments. They typically contain MCS, selection markers, and high-copy-number origins.

- **Expression Vectors** are used to produce proteins in host cells and include strong promoters (e.g., T7, lac, or CMV), ribosome binding sites, and regulatory elements to control expression levels.
- **Metabolic Engineering Vectors** are tailored to introduce entire biosynthetic pathways. These often carry multiple genes under separate or polycistronic promoters and may include regulatory circuits for pathway optimization.
- **Therapeutic Vectors** (e.g., viral vectors like lentivirus or AAV) are employed for gene therapy and must be safe, efficient in delivery, and capable of long-term gene expression in human or animal cells. Such vectors are often subject to rigorous safety and regulatory standards.

For example, Gateway® technology and Golden Gate Assembly enable high-throughput construction of expression systems in fungi and bacteria for applications in biomanufacturing and drug discovery [35].

13.5.3 Mechanisms of Waste Degradation by GEMs

GEMs utilize an array of advanced biochemical pathways to break down, neutralize, or convert a wide variety of waste substances. These pathways are specifically modified through genetic engineering to improve the effectiveness, selectivity, and long-term sustainability of microbial degradation processes. This enhancement allows GEMs to efficiently target and treat diverse waste categories such as heavy metals, synthetic plastics, petroleum-based hydrocarbons, industrial dyes, and numerous other organic pollutants. GEMs perform waste degradation through various biochemical mechanisms:

- **Enzymatic Hydrolysis and Oxidation:** One of the primary mechanisms used by GEMs involves the overexpression and overproduce of degradative enzymes like oxygenases, dehalogenases, hydrolases, and laccases to break down complex pollutants [36]. These enzymes catalyze the cleavage of complex or recalcitrant pollutants into simpler, less toxic molecules. For example, engineered *Pseudomonas putida* strains have been modified to overexpress monooxygenases and dioxygenases that effectively degrade aromatic hydrocarbons in petrochemical waste [37]. Similarly, fungi like *Aspergillus niger*, engineered to produce high levels of laccase and manganese peroxidase, exhibit enhanced degradation of lignin and industrial dyes in wastewater [38].
- **Biosorption and Bioaccumulation:** Metal-binding proteins or metallothioneins can be expressed for heavy metal uptake [39]. GEMs are also engineered to express metal-binding proteins, such as metallothioneins, phytochelatins, or surface-displayed peptides, which facilitate the biosorption and bioaccumulation of heavy metals from contaminated environments. For instance, *Bacillus subtilis* engineered with *smtA*, a gene encoding metallothionein, has demonstrated superior ability to bind and immobilize cadmium and lead from industrial effluents [40]. These mechanisms are particularly vital for the treatment of electronic and industrial waste streams containing high concentrations of heavy metals.

- **Synthetic Gene Circuits and Novel Pathways:** Advanced synthetic biology allows the creation of novel pathways for the complete mineralization of toxic compounds [41]. These advances have enabled the design of synthetic gene circuits and custom metabolic pathways that allow GEMs to process pollutants that are otherwise recalcitrant to natural microbial degradation. These circuits can regulate the sequential expression of multiple enzymes, enabling the complete mineralization of complex compounds such as polychlorinated biphenyls (PCBs), endocrine-disrupting chemicals, or mixed solvents [42]. Engineered *E. coli* strains with modular biosynthetic pathways, for example, can degrade herbicides and pharmaceuticals through combinatorial biotransformation steps [43].
- **Co-metabolism:** GEMs can also utilize a process known as co-metabolism, where they degrade pollutants incidentally while metabolizing an auxiliary carbon source [44]. In this process, the enzymes produced for the primary metabolism act nonspecifically on the pollutant. This mechanism is particularly useful for degrading xenobiotics that cannot support microbial growth alone. For example, GEMs engineered from *Pseudomonas fluorescens* can degrade TCE while growing on lactate or methane as a primary substrate [45]. This dual-pathway strategy enables effective treatment of mixed industrial or municipal waste streams.

13.5.4 GEMs for Waste Treatment

The integration of microbiological techniques, molecular biology, ecological understanding, and biochemical pathways is thoroughly examined for the development of GEMs that accelerate the biodegradation process. The incorporation of additional genetic traits into these microorganisms significantly enhances their ability to break down and remove environmental pollutants more effectively [8].

In the process of designing GEMs, a distinct plasmid is necessary for the degradation of each specific compound, as a single plasmid is not capable of breaking down all toxic substances that belong to different chemical groups. These plasmids are classified into four main categories based on the compounds they target: the OCT plasmid is responsible for degrading hydrocarbons such as octane, hexane, and decane; the XYL plasmid facilitates the breakdown of aromatic compounds like xylene and toluene; the CAM plasmid is involved in the decomposition of camphor; and the NAH plasmid specializes in degrading polycyclic aromatic hydrocarbons such as naphthalene [46]. *Pseudomonas putida* harbors both the XYL and NAH plasmids, along with a hybrid plasmid formed through the recombination of segments from the CAM and OCT plasmids. This combination enables the bacterium to effectively degrade a range of compounds, including camphor, naphthalene, octane, and salicylate [11]. This multiplasmid-containing *Pseudomonas* strain was also having a great potential to metabolizing hydrocarbons of different types such as aliphatic, aromatic, terpenic, and polyaromatic hydrocarbons (PAHs). It could grow rapidly on crude oil, and so the strain was more efficient than any other single plasmid [46].

Pseudomonas putida is considered a highly promising strain for applications in metabolic engineering and genetic modification. This microorganism has the capability to express genes that encode various degradative enzymes. Through genetic engineering, *P. putida* has been modified to enhance its efficiency in degrading environmental pollutants such as naphthalene and salicylate [47]. The genetically engineered organism known as the "superbug" or "oil-eating bug" is a product of advanced genetic modification. It was developed using various plasmids derived from *P. putida*, including TOL (which enables degradation of toluene and xylene), RA500 (responsible for breaking down 3,5-xylene), pAC25 (for degrading 3-chloro-benzoate), and pKF439 (which aids in the conversion of salicylate and toluene). These plasmids allow the engineered microorganism to degrade a wide range of chemical pollutants. The WWO plasmid originally identified in *P. putida* is now recognized as the TOL plasmid. Notably, this "superbug" was the first living organism to become the subject of a legal case involving intellectual property rights [11].

During the early 1980s, GMMs were created using emerging genetic engineering technologies alongside extensive research into the metabolic capabilities of various microorganisms. This scientific progress led to the development of the first two genetically modified bacterial strains, *Pseudomonas aeruginosa* (NRRL B-5472) and *P. putida* (NRRL B-5473), both of which were officially patented in the United States in the year 1981. In the early 1970s, Chakrabarty constructed these GMMs from genes for naphthalene, salicylate, and camphor degradation [48].

A joint initiative between the University of Tennessee and Oak Ridge National Laboratory successfully led to the development of the first genetically engineered strain, *P. fluorescens* HK44, specifically designed for the bioremediation of soil contaminated with naphthalene [49, 50]. The parent strain used to develop *P. fluorescens* HK44 was originally isolated from a manufactured gas plant site that was significantly polluted with polyaromatic hydrocarbons (PAHs). The engineered *P. fluorescens* HK44 harbored a naphthalene-degrading catabolic plasmid, designated as pUTK21, which was constructed by inserting transposon Tn4431 into the NAH7 plasmid derived from *P. fluorescens* strain 5R [51, 52]. This HK44 strain also carried a transposon-based bioluminescent lux gene, which was genetically fused to a promoter associated with the naphthalene catabolic genes, enabling the strain to produce bioluminescence in response to the presence of naphthalene [53].

13.6 Applications of GEMs in Waste Treatment

13.6.1 Industrial Waste

GEMs have demonstrated considerable potential in the bioremediation of various industrial waste streams, including those contaminated with petroleum hydrocarbons, heavy metals, and pharmaceutical residues. A notable example is *P. putida*, which has been genetically altered to enhance its capacity to break down aromatic hydrocarbons such as toluene and xylene more efficiently [54]. Engineered strains of *E. coli* have also been employed to degrade phenolic compounds in textile effluents (Table 13.2) [55].

Table 13.2 Genetically modified microorganisms for biodegradation of industrial waste.

S. No.	Genetically modified microorganism	Engineered gene(s)	Mode of action	Targeted industrial waste
1	*Pseudomonas putida* KT2440 [37]	*todC1C2BA, xylE*	Degradation of aromatic hydrocarbons and toluene	Petrochemical effluent
2	*Escherichia coli* JM109 [56]	*nahA, nahB* from *Pseudomonas* sp.	Converts naphthalene into salicylate and catechol	PAHs in refinery waste
3	*Bacillus subtilis* [57]	*lipA, cutinase*	Breaks down industrial synthetic esters and plasticizers	Plasticizer-rich plastic processing waste
4	*Cupriavidus necator* [58]	*phaCAB* genes for PHA biosynthesis	Converts carbon-rich effluent into bioplastics	Waste glycerol from biodiesel plants
5	*Rhodococcus erythropolis* [59]	*alkB, rubA*	Hydrocarbon monooxygenase system degrades alkanes	Oil spills and chemical manufacturing runoff
6	*Aspergillus niger* [38]	Overexpression of *MnP* and *LiP* (peroxidases)	Degradation of lignin and dye residues	Pulp and paper mill waste
7	*Clostridium butyricum* [60]	Engineered butyrate production pathway (*thl, buk*)	Fermentation of organic waste into butyric acid	Organic residues from food industries
8	*Yarrowia lipolytica* [61]	*POX2, LIP2, CYP52*	Biodegradation of long-chain fatty acids	Grease trap and oil processing waste
9	*Alcaligenes eutrophus* [62]	*pcbC* for PCB dechlorination	Biotransformation of polychlorinated biphenyls	Electronic chemical industry effluent
10	*Corynebacterium glutamicum* [63]	*pcaHGBDC* (protocatechuate pathway)	Aromatic compound degradation from dye intermediates	Dye and pigment industry wastewater

Industrial waste often contains hazardous chemicals, including petrochemicals, aromatic compounds, dyes, oils, plasticizers, heavy metals, and recalcitrant organics. GEMs can:

- Break down complex organics (e.g., PAHs, alkanes, phenols).
- Convert toxic waste into value-added products like bioplastics or bioacids.
- Aid in bioremediation of effluents from industries such as petrochemical, dye, paper, biodiesel, and plastics.

13.6.2 Agricultural Waste

In agriculture, GEMs help in breaking down pesticide residues and converting crop residues into value-added products. Modified strains of *B. subtilis* have been used for organophosphate degradation, while certain engineered Rhizobium species help in nitrogen fixation and soil fertility enhancement (Table 13.3) [64].

These GEMs significantly improve the biodegradation efficiency of complex agricultural wastes such as rice straw, corn stover, wheat bran, sugarcane bagasse, and

Table 13.3 Genetically modified microorganisms for biodegradation of agricultural waste.

S. No.	Genetically modified microorganism	Engineered gene(s)	Mode of action	Targeted agricultural waste
1	*Trichoderma reesei* [65]	Overexpression of *cbh1, egl1, xyn2* (cellulases and xylanases)	Enhanced lignocellulose degradation	Sugarcane bagasse, rice straw
2	*Bacillus subtilis* [66]	*xynA, eglS* (xylanase and endoglucanase)	Synergistic hydrolysis of hemicellulose and cellulose	Wheat straw, corn stover
3	*Escherichia coli (engineered for ethanol production)* [67]	*celA, xynB* from *Clostridium thermocellum*	Converts cellulose and xylan to fermentable sugars and ethanol	Rice husk, corn stalk
4	*Pseudomonas putida KT2440* [68]	*vanAB, ech* for lignin degradation	Catabolizes aromatic lignin-derived compounds	Wheat bran, peanut shells
5	*Saccharomyces cerevisiae* [69]	*xyl1, xyl2, xyl3* (xylose utilization genes)	Enables fermentation of xylose-rich hydrolysates	Corn cob hydrolysate
6	*Clostridium thermocellum* [70]	Enhanced expression of *cipA, celS* (cellulosome scaffoldins)	Direct hydrolysis of crystalline cellulose	Paddy straw, corn stalk
7	*Aspergillus niger* [71]	Overexpressed *lacA* (laccase gene)	Oxidizes lignin for improved cellulose access	Wheat straw, cotton stalk
8	*Streptomyces coelicolor* [72]	Introduction of *LiP (lignin peroxidase)* and *MnP* (manganese peroxidase)	Degrades lignin-derived aromatic structures	Rice straw, sugarcane trash
9	*Zymomonas mobilis* [73]	*bglB* (β-glucosidase gene) from *Thermotoga maritima*	Converts cellobiose to glucose for ethanol fermentation	Sorghum bagasse
10	*Rhizopus oryzae* [74]	Expression of fungal xylanases and cellulases	Degrades hemicellulose in agro-residues	Barley husk, millet waste

lignocellulosic residues. They do this through the overexpression of key cellulolytic, hemicellulolytic, or lignin-degrading enzymes or by enabling fermentation pathways for biofuel or value-added product generation.

13.6.3 Municipal Solid Waste

The organic fraction of Municipal Solid Waste (MSW) offers an excellent substrate for GEM-based treatment. Engineered anaerobic bacteria such as *Clostridium thermocellum* have shown increased lignocellulose degradation, thereby enhancing methane yield during anaerobic digestion [75]. Engineered methanogens further improve biogas production from municipal waste (Table 13.4) [76].

Table 13.4 Genetically modified microorganisms for biodegradation of municipal solid waste (MSW).

S. No.	Genetically modified microorganism	Engineered gene(s)	Mode of action	Targeted MSW components
1	*Escherichia coli* BL21 (DE3) [77]	*phaCAB* from *Ralstonia eutropha*	Converts organic fraction of MSW into polyhydroxyalkanoates (PHAs)	Biodegradable food waste, sugars
2	*Bacillus subtilis* 168 [78]	*cotA* (laccase gene)	Oxidative breakdown of dyes and synthetic polymers	Textile waste in MSW
3	*Pseudomonas putida* KT2440 [79]	Overexpressed *alkB* and *phaC1*	Degrades aliphatic hydrocarbons and converts into bioplastics	Plastic waste residues
4	*Saccharomyces cerevisiae* [80]	*ADH2, ALD6, ATF1*	Converts fermentable sugars into bioethanol	Organic kitchen waste
5	*Clostridium thermocellum* [81]	*celS, cipA*	Hydrolyzes cellulose and hemicellulose from compostable MSW	Paper, cardboard, plant biomass
6	*Aspergillus niger* [82]	Overexpression of *cutinase, esterase*	Biodegradation of PET and packaging waste	Synthetic polymer waste
7	*Cupriavidus necator* [83]	*phaC, phaA, phaB*	Converts volatile fatty acids into PHAs	Organic fractions of MSW
8	*Lactobacillus plantarum* [84]	*ldhL, ldhD* for D-/L-lactate	Produces lactic acid from food waste for PLA synthesis	Food scraps, fruit peel
9	*Rhizopus oryzae* [85]	*amyA* (amylase gene)	Converts starch-rich waste into organic acids	Cooked food residues
10	*Streptomyces lividans* [86]	*xylE, catA*	Breaks down aromatic hydrocarbons from urban MSW	Solvent-contaminated plastics and containers

MSW contains complex, heterogeneous materials including food waste, plastics, paper, textile waste, and hazardous organics. GEMs have shown significant promise in enhancing the degradation and valorization of these waste fractions. Their metabolic pathways are specifically engineered to transform waste materials into useful end products such as biofuels, bioplastics, organic acids, and various other high-value compounds.

13.6.4 E-waste and Heavy Metal Contamination

Certain GEMs are deliberately designed to bioaccumulate or detoxify specific heavy metals, including arsenic, cadmium, copper, cobalt, lead, mercury, nickel, and zinc, making them highly effective for applications in environmental cleanup and pollution control. Genetically modified *Ralstonia metallidurans* and *Cupriavidus metallidurans* are capable of sequestering and transforming heavy metals into less toxic forms [87]. The compilation of GEMs developed for the desorption of heavy metal ions is presented in Tables 13.5 and 13.6.

E-waste, composed of discarded electronics like printed circuit boards (PCBs), contains precious metals (e.g., gold, silver, copper) and hazardous materials (e.g., lead, cadmium, arsenic). GEMs offer an eco-friendly alternative to chemical leaching by:

- Enhancing bioleaching through organic acid production.
- Improving metal efflux, adsorption, and detoxification.
- Enabling biotransformation of metals into recoverable forms.
- Supporting metal-tolerant growth under toxic conditions.

Table 13.5 Genetically modified microorganisms for biodegradation and biorecovery of e-waste.

S. No.	Genetically modified microorganism	Engineered gene(s)	Mode of action	Targeted e-waste component
1	*Pseudomonas aeruginosa* [88]	Overexpression of *copA, cueO*	Enhanced copper efflux and oxidation	Copper recovery from PCBs
2	*Ralstonia metallidurans* CH34 [89]	*czcABC, cop* operon	Multimetal resistance and efflux	Cadmium, zinc, copper from circuit boards
3	*Escherichia coli* DH5α [90]	*metH, arsC*	Bioaccumulation of gold and reduction of arsenic	Gold recovery, arsenic detoxification
4	*Aspergillus niger* [91]	Overexpression of *citA* (citrate synthase)	Bioleaching via increased citric acid production	Leaching of Cu, Zn, Ni from shredded boards

S. No.	Genetically modified microorganism	Engineered gene(s)	Mode of action	Targeted e-waste component
5	*Shewanella oneidensis* MR-1 [92]	Overexpression of *mtrCAB* complex	Extracellular electron transfer mechanisms play a critical role in the microbial reduction of metals	Palladium and copper reduction
6	*Saccharomyces cerevisiae* [93]	Expression of metallothioneins (*CUP1, Crs5p*)	Bioadsorption and binding of precious metals	Silver and copper ions
7	*Cupriavidus necator* [94]	Engineered for polyhydroxybutyrate (PHB) production	Bioplastic production using organic e-waste leachate	Printed circuit board (PCB) leachate
8	*Bacillus subtilis* [40]	*smtA* (metallothionein gene)	Metal chelation and detoxification	Lead and cadmium from soldering waste
9	*Corynebacterium glutamicum* [95]	*mntH* (Mn transporter), *copA*	Enhanced metal ion uptake systems	Recovery of rare earth elements (REEs)
10	*Yarrowia lipolytica* [96]	Overexpression of *LIP2* and acid-producing genes	Organic acid-mediated bioleaching	Ni, Zn, Cr from e-waste fines

Table 13.6 Comprehensive list of genetically modified microorganisms for the desorption of heavy metal.

S. No.	Genetically modified microorganisms	Introduced gene(s)	Mode of action/ expression	Targeted heavy metal(s)
1	*Deinococcus radiodurans* [97]	*merH* (ion transporter)	Enhances mercury resistance and transforms mercury ions	Hg^{2+}
2	*Acidithiobacillus ferrooxidans* [98]	*merC*	Facilitates mercury uptake and detoxification	Hg^{2+}
3	*Mesorhizobium huakuii* [99]	Genes from *Arabidopsis thaliana* (phytochelatin synthase)	Accumulates and detoxifies cadmium	Cd^{2+}

(Continued)

Table 13.6 (Continued)

S. No.	Genetically modified microorganisms	Introduced gene(s)	Mode of action/ expression	Targeted heavy metal(s)
4	*Escherichia coli* SE5000 [100]	*nixA*	Induces nickel-specific transporters	Ni^{2+}
5	*Pseudomonas* K-62 [101]	Organomercurial lyase and reductase	Breaks down methyl mercury and inorganic Hg	Hg^{2+}, CH_3Hg^+
6	*Ralstonia eutropha* AUM-01 [102]	Engineered rhizobial plasmids	Binds and accumulates multiple metals	Mixed metals
7	*Rhodopseudomonas palustris* [103]	Mercury transporter genes	Detoxifies and converts mercury to elemental form	Hg^{2+}
8	*Pseudomonas putida* X3 [104]	*EGFP*, methyl parathion and cadmium degradation genes	Degrades Cd and methyl parathion	Cd^{2+}, methyl parathion
9	*Cupriavidus metallidurans* [105] CH34	*czcABC, copA, mer* genes	Metal efflux systems; resistance to Cd, Zn, Co, Hg	Cd^{2+}, Zn^{2+}, Co^{2+}, Hg^{2+}
10	*Bacillus subtilis* BSN29 [106]	Overexpression of *smtA* (metallothionein gene)	Metal binding via cysteine-rich proteins	Pb^{2+}, Cd^{2+}
11	*Synechocystis* sp. PCC 6803 [107]	*arsB, arsC* (arsenic resistance genes)	Biotransformation of arsenate to arsenite and efflux	As^{5+}, As^{3+}
12	*Saccharomyces cerevisiae* [108]	*CUP1* (copper-binding protein)	Biosorption of copper ions via cell wall proteins	Cu^{2+}
13	*Shewanella oneidensis* MR-1 [109]	Overexpression of metal reductase genes	Reductive transformation of Cr (VI) to Cr (III)	Cr^{6+}

13.7 Advantages of Using GEMs

The use of GEMs in waste management and environmental cleanup presents several significant advantages when compared to traditional biological and chemical treatment methods. Due to their enhanced genetic makeup, GEMs can be fine-tuned for higher efficiency, specificity, and sustainability, making them a powerful tool in modern waste management strategies.

13.7.1 High Efficiency and Performance Under Variable Conditions

GEMs show superior degradation performance under diverse environmental conditions. This implies that they are engineered to retain strong degradation efficiency even when exposed to variable or challenging environmental conditions, including extreme levels of pH, temperature, salinity, or high concentrations of contaminants [110]. For instance, engineered strains of *Deinococcus radiodurans* and *P. putida* have demonstrated efficient degradation of heavy metals and hydrocarbons, respectively, in toxic and radioactive environments where native microbes fail to survive [39, 97]. The ability to overexpress enzymes like oxygenases, laccases, or dehalogenases contributes significantly to the high degradation efficiency of GEMs across varied waste types.

13.7.2 Targeted and Specific Action

Engineered microbes can be customized for specific types of waste or pollutants. One of the most powerful advantages of GEMs lies in their **programmable specificity**. Through synthetic biology and gene editing tools such as CRISPR-Cas9, microbes can be engineered to express **specific catabolic enzymes** that target a particular class of pollutants—such as endocrine disruptors, chlorinated solvents, or pharmaceutical residues [41]. This precision reduces unintended side effects on nontarget compounds and ensures selective degradation, which is crucial in the treatment of complex waste mixtures such as e-waste, industrial sludge, or MSW.

13.7.3 Economic Viability and Resource Recovery

GEMs help to reduce energy inputs, and improved recovery of bio-based products lowers treatment costs. GEMs contribute to cost-effective waste treatment by enabling simultaneous degradation and valorization of waste. For example, engineered *Cupriavidus necator* can convert organic fractions of industrial effluent into polyhydroxyalkanoates (PHAs), a type of biodegradable plastic, while *E. coli* modified with ethanol production pathways can ferment municipal food waste into bioethanol [58, 80]. This dual functionality waste treatment combined with product recovery significantly lowers operational costs by reducing energy inputs and generating commercially valuable by-products.

13.7.4 Environmental Sustainability

GEMs minimize secondary pollution and reduce greenhouse gas emissions when integrated into circular waste management systems. Incorporating GEMs into waste treatment processes aligns with environmentally sustainable and circular economy principles [111]. Unlike chemical methods that may result in toxic residues or greenhouse gas emissions, GEMs offer greener alternatives by operating under mild conditions and minimizing secondary pollution [78]. Moreover, GEM-based systems can be integrated into closed-loop waste management frameworks, where waste is not only detoxified but also converted into useful products, reducing the environmental footprint and conserving natural resources.

13.8 Challenges and Concerns

While GEMs offer innovative solutions for environmental waste treatment, their application is accompanied by several significant challenges and concerns despite their potential. GEMs raise several challenges that are follows:

13.8.1 Ecological Risks

One of the foremost concerns is the potential for ecological disruption. There is concern about horizontal gene transfer and unintended ecosystem interactions [112]. GEMs may transfer genetic material to native microorganisms through horizontal gene transfer, potentially leading to the emergence of new, unintended microbial traits in the environment [100, 107]. This raises fears of creating antibiotic resistance or altering microbial community structures, which could destabilize existing ecosystems.

13.8.2 Regulatory Restrictions

The environmental release of GEMs is tightly regulated and requires thorough risk assessments [113]. The release of GEMs into the environment is highly regulated due to these potential risks. National and international frameworks, including biosafety protocols, require extensive risk assessment and approval procedures before field deployment [98, 106]. These regulatory hurdles can significantly delay the practical application of GEM-based technologies.

13.8.3 Limited Field Data

Most research on GEMs remains confined to controlled laboratory conditions or pilot-scale experiments studies and their field efficacy remains under-researched [114]. There is a lack of comprehensive field studies that can validate their efficiency, stability, and safety under real-world environmental conditions [97, 99]. This gap in field data makes it difficult to generalize laboratory findings or ensure consistent performance across diverse ecological settings.

13.8.4 Public Acceptance

Ethical and safety concerns about genetically modified organisms (GMOs) may hinder large-scale implementation [115]. Public Acceptance: Ethical, cultural, and safety concerns surrounding GMOs remain a barrier to public acceptance of GEMs. Public resistance is often driven by a lack of awareness or misinformation about the risks and benefits, which can hamper large-scale implementation and funding for such technologies [103, 109].

Overall, addressing these challenges requires interdisciplinary collaboration among scientists, policymakers, and the public to ensure that GEMs are developed and applied responsibly.

13.9 Biosafety and Regulation

The release and use of GEMs are subject to strict national and international guidelines. The Cartagena Protocol on Biosafety outlines safe handling procedures for GMOs. Containment mechanisms such as auxotrophic mutations, kill-switches, and genetic firewall systems are essential to mitigate risks associated with GEMs in the environment [116]. Regulatory frameworks in countries like the USA, EU, and India also mandate biosafety assessments before deployment.

13.10 Future Prospects and Innovations

With advances in omics technologies and AI-assisted synthetic biology, the future of GEMs in waste management looks promising:

- Development of engineered consortia to target complex waste mixtures.
- Use of biosensors within GEMs to monitor pollutant levels in real time.
- Application in mobile bioreactors for on-site waste treatment in rural and remote areas.
- Integration with machine learning models to predict optimal metabolic pathways for new pollutant types [117].

13.11 Conclusion

GEMs represent a groundbreaking approach to modern waste treatment. Their high adaptability, operational efficiency, and potential for recovering valuable resources position them as essential components of future-oriented waste management systems. Despite ongoing challenges related to safety concerns, regulatory frameworks, and public acceptance, continual progress in genetic engineering and environmental biotechnology are expected to overcome these obstacles. When responsibly implemented, GEMs can serve as key drivers in the transition toward a sustainable, circular, and environmentally friendly bioeconomy.

At the heart of genetic engineering processes lie plasmids and vectors. Ranging from traditional plasmid systems to advanced artificial chromosomes and CRISPR-compatible constructs, these genetic tools form the foundation for constructing GEMs aimed at environmental remediation, industrial-scale bioprocessing, and biomedical innovations. As synthetic biology continues to evolve, the precision, scalability, and versatility of vector platforms are also expanding, enabling the exploration of new and more complex applications within microbial biotechnology.

Modern vector systems have undergone significant transformation through the incorporation of fusion tags, shuttle vector capabilities, and CRISPR compatibility. These technological enhancements have turned vectors into robust, modular platforms that support advanced functions, including therapeutic gene editing, sophisticated metabolic engineering, and real-time visualization of cellular activities. Selecting an appropriate vector is critical for improving the accuracy, efficiency, and reproducibility of genetic engineering techniques. To achieve optimal outcomes in scientific research, industrial biotechnology, or therapeutic development, the vector must be carefully matched to factors such as the DNA insert size, the host organism, and the specific objectives of the experiment.

References

1 Ghosh, A. (2019). Genetically engineered microbes for environmental remediation. *Biotechnol Adv.* 37 (6): 107440.

2 Rafeeq, H., Afsheen, N., Rafique, S., et al. (2023). Genetically engineered microorganisms for environmental remediation. *Chemosphere* 310: 136751.

3 Venulet, J. and Van Etten, R.L. (1970). *The Chemistry of the Nitro and Nitroso Groups* (ed. H. Fever), 201–289. New York: Interscience.

4 Vargas, J.M. (1975). Pesticide degradation. *J Arboricult.* 1 (12): 232–233.

5 Chawla, N., Suneja, S., Kukreja, K., and Kumar, R. (2013). Bioremediation: An emerging technology for remediation of pesticides. *Res J Chem Environ.* 17 (4): 88–105.

6 Singh, R. (2020). CRISPR/Cas9: a robust genome editing tool for bioremediation. *Bioresour Technol.* 307: 123123.

7 Kiran, S. (2021). Role of engineered microbes in industrial waste degradation. *Environ Chem Lett.* 19 (1): 505–518.

8 Cases, I. and de Lorenzo, V. (2005). Genetically modified organisms for the environment: stories of success and failure and what we have learned from them. *Int Microbiol.* 8 (3): 213–222.

9 Mrozik, A., Piotrowska-Seget, Z., and Łabużek, S. (2005). Bacteria in bioremediation of hydrocarbon-contaminated environments. *Microbiol Post.* 44 (3): 227–238.

10 Garg, M. (2016). Xenobiotic compounds: meaning, hazards and biodegradation. www.biologydiscussion.com/microbiology2/bioremediation/xenobiotic compounds-meaning-hazards-and-biodegradation/55625 (accessed 03 July 2019).

11 Sayler, G.S. and Ripp, S. (2000). Field applications of genetically engineered microorganisms for bioremediation processes. *Curr Opin Biotechnol.* 11: 286–289.

12 Ramos, J.L., Díaz, E., Dowling, D., et al. (1994). The behavior of bacteria designed for biodegradation. *Biotechnology.* 12 (13): 1349–1356.

13 Dale, J.W. and Park, S.F. (2007). *Molecular Genetics of Bacteria*, 237–244. University of Surrey.

14 Kunicki-Goldfinger, W. (2007). *Życie bakterii*, 267–344. PWN.

15 Bansal, A.K. (2005). Bioinformatics in microbial biotechnology – a mini review. *Microbial Cell Factor.* 4: 19.

16 Menn, F.M., Easter, J.P., and Sayler, G.S. (2008). Genetically engineered micro-organisms and bioremediation. *Biotechnol: Environ Process.* 11(b): 29–37.

17 Chmiel, A. (1998). *Biotechnologia – Podstawybiotechnologii*, 260–306. PWN.

18 Nair, A.J. (2008). *Introduction to Biotechnology and Genetic Engineering*, 467–776. Infinity Science Press LLC.

19 Watson, J.D., Baker, T.A., Bell, S.P., et al. (2004). *Molecular Biology of the Gene*, 293–342. Pearson Education Inc.

20 Sambrook, J. and Russell, D.W. (2001). *Molecular Cloning: A Laboratory Manual*, 3e. Cold Spring Harbor Laboratory Press.

21 Brown, T.A. (2010). *Gene Cloning and DNA Analysis: An Introduction*, 6e. Wiley-Blackwell.

22 Ausubel, F.M., Brent, R., Kingston, R.E., et al. (2002). *Short Protocols in Molecular Biology*, 5e. Wiley.

23 Green, M.R. and Sambrook, J. (2012). *Molecular Cloning: A Laboratory Manual*, 4e. Cold Spring Harbor Laboratory Press.

24 Nielsen, J. and Keasling, J.D. (2016). Engineering cellular metabolism. *Cell.* 164 (6): 1185–1197. https://doi.org/10.1016/j.cell.2016.02.004.

25 Brown, T.A. (2016). *Gene Cloning and DNA Analysis: An Introduction*, 7e. Wiley-Blackwell.

26 Davison, J. (2005). Risk mitigation of genetically modified bacteria and plants de-signed for bioremediation. *J Indus Microbiol Biotechnol.* 32 (11–12): 639–650.

27 Sayler, G.S., Matrubutham, U., Menn, F.M., et al. (1998). Molecular probes and biosensors in bioremediation and site assessment. In: *Bioremediation: Principles and Practice* (ed. S.K. Sikdar), 1e, 385–434. Routledge.

28 Villacieros, M., Whelan, C., Mackova, M., et al. (2005). Polychlorinated biphenyl rhizore-mediation by *Pseudomonas fluorescens* F113 derivatives, using a *Sinorhizobium meliloti nod* system to drive bph gene expression. *Appl Environ Microbiol.* 71 (5): 2687–2694.

29 Maniatis, T., Fritsch, E.F., and Sambrook, J. (1982). *Molecular Cloning: A Laboratory Manual.* Cold Spring Harbor Laboratory.

30 Zhang, Y., Li, X., Wang, Y., et al. (2023). CRISPR-based synthetic circuits for microbial cell engineering. *Nat Commun.* 14 (1): 1256.

31 Liu, W., Zheng, F., Chen, W., et al. (2022). Engineering plasmid vectors for dynamic control of gene expression in *E. coli. Biotechnol J.* 17 (5): e2100598.

32 Watson, J.D., Gilman, M., Witkowski, J., and Zoller, M. (2007). *Recombinant DNA: Genes and Genomes – A Short Course*, 3e. W. H. Freeman.

33 Osoegawa, K., Tateno, M., Woon, P.Y., et al. (2000). BAC clones: a vital tool for genomics research. *Genomics.* 65 (1): 1–13.

34 Meyer, V., Arentshorst, M., El-Ghezal, A., et al. (2007). Highly efficient gene targeting in the *Aspergillus niger kusA* mutant. *J Biotechnol.* 128 (4): 770–775.

35 Liu, Q., Gao, R., Li, J., and Lin, Z. (2023). Efficient pathway assembly in Saccharomyces cerevisiae using modular Golden Gate constructs. *FEMS Yeast Res.* 23 (1): foad004.

36 Narayanan, M. (2019). Enzymatic degradation of persistent pollutants: role of engineered biocatalysts. *Appl Microbiol Biotechnol.* 103 (4): 1425–1437.
37 Nikel, P.I. (2016). Pseudomonas putida as a functional chassis for industrial biocatalysis: from native to synthetic metabolism. *Nat Chem Biol.* 12 (9): 755–763.
38 Sharma, B. (2019). Enhanced ligninolytic enzyme production in genetically modified *Aspergillus niger* for industrial dye degradation. *Environ Technol Innov.* 13: 263–271.
39 Srivastava, S. (2018). Engineered microbes for heavy metal biosorption. *J Appl Microbiol.* 124 (2): 313–327.
40 Kumar, R. and Yadav, A. (2021). Genetic engineering of Bacillus subtilis for heavy metal bioremediation using metallothioneins. *Ecotoxicol Environ Saf.* 214: 112134.
41 Brophy, J.A. and Voigt, C.A. (2014). Principles of genetic circuit design. *Nat Methods.* 11 (5): 508–520.
42 Chandran, S. (2022). Synthetic gene circuits for pollutant mineralization in genetically engineered microorganisms. *Trends Biotechnol.* 40 (2): 125–138.
43 Liang, C. (2020). Modular biosynthetic pathways in *E. coli* for degradation of pharmaceutical residues. *Biotechnol Adv.* 44: 107623.
44 Pandey, R. (2020). Co-metabolic degradation of complex pollutants by GEMs. *Biodegradation.* 31 (1–2): 45–58.
45 Semprini, L. (2013). Cometabolic biodegradation of chlorinated solvents using engineered bacteria. *J Hazard Mater.* 261: 83–90.
46 Markandey, D.K. and Rajvaidya, N. (2004). *Environmental Biotechnology*, 1e, 79. APH Publishing Corporation.
47 Samanta, S.K., Bhushan, B., and Jain, R.K. (2001). Efficiency of naphthalene and salicylate degradation by a recombinant *Pseudomonas putida* mutant strain defective in glucose metabolism. *Appl Microbiol Biotechnol.* 55: 627–631.
48 Wasilkowski, D., Swędzioł, Ż., and Mrozik, A. (2012). The applicability of genetically modified microorganisms in bioremediation of contaminated environments. *Chemik.* 66 (8): 817–826.
49 Ripp, S., Nivens, D.E., Ahn, Y., et al. (2000). Controlled field release of a bioluminescent genetically engineered microorganism for bioremediation process monitoring and control. *Environ Sci Technol.* 34 (5): 846–853.
50 Sayler, G.S., Cox, C.D., Burlage, R., et al. (1999). Field application of a genetically engineered microorganism for polycyclic aromatic hydrocarbon bioremediation process monitoring and control. In: *Novel Approaches for Bioremediation of Organic Pollution*, 241–254. New York: Kluwer Academic/Plenum Publishers.
51 Vandevivere, P.C., Bianchi, R., and Verstraete, W. (1999). Treatment and reuse of wastewater from the textile wet-processing industry: review of emerging technologies. *J Chem Technol Biotechnol.* 72: 289–302.
52 King, J.M.H., Digrazia, P.M., Applegate, B., et al. (1990). Rapid, sensitive bioluminescence reporter technology for naphthalene exposure and biodegradation. *Science.* 249 (4970): 778–781.
53 Chatterjee, J. and Meighen, E.A. (2008). Biotechnological applications of bacterial bioluminescence (*lux*) genes. *Photochem Photobiol.* 62 (4): 641–650.

54 Ramos, J.L. (2016). Engineering *Pseudomonas putida* for efficient degradation of aromatic hydrocarbons. *Nat Rev Microbiol.* 14 (11): 755–764.

55 Kumar, R. (2018). Textile dye biodegradation by modified *E. coli. J Hazard Mater.* 360: 173–180.

56 Ahmed, M. (2019). Genetically modified *E. coli* for naphthalene degradation in refinery waste. *Bioresour Technol.* 281: 330–338.

57 Kumar, V. and Das, S. (2020). Enzymatic plasticizer degradation using *Bacillus subtilis. J Hazard Mater.* 388: 121735.

58 Khosravi-Darani, K. (2018). Bioplastic synthesis from biodiesel waste by *C. necator. Int J Biol Macromol.* 116: 535–545.

59 de Carvalho, C.C.C.R. (2020). Hydrocarbon degradation by *Rhodococcus erythropolis. J Ind Microbiol Biotechnol.* 47 (5–6): 383–395.

60 Wang, Y. (2020). Fermentative valorization of food waste using engineered *C. butyricum. Bioresour Technol.* 298: 122554.

61 Abdel-Mawgoud, A.M. (2018). Lipid degradation by *Yarrowia lipolytica* for industrial waste treatment. *Microb Cell Fact.* 17 (1): 142.

62 Field, J.A. (2021). Biotransformation of PCBs using *A. eutrophus. Biotechnol Adv.* 50: 107767.

63 Becker, J. (2021). Aromatic pollutant degradation by *Corynebacterium glutamicum. Biotechnol J.* 16 (6): e2000445.

64 Wang, J. (2020). Biodegradation of organophosphates using genetically modified *Bacillus* strains. *J Environ Manag.* 262: 110315.

65 Li, X. (2021). Overexpression of lignocellulolytic enzymes in *Trichoderma reesei* for efficient biomass degradation. *Bioresour Technol.* 320: 124422.

66 Sharma, R. (2020). Genetic engineering of *Bacillus subtilis* for improved degradation of crop residues. *Ind Crops Prod.* 148: 112314.

67 Kim, S.R. (2019). Engineered *E. coli* expressing cellulase and xylanase for consolidated bioprocessing. *Biotechnol Bioeng.* 116 (1): 87–96.

68 Salvachúa, D. (2015). *Pseudomonas putida* KT2440 engineered for lignin valorization. *Appl Environ Microbiol.* 81 (15): 5312–5324.

69 Almeida, J.R. (2011). Engineering *S. cerevisiae* for xylose fermentation. *FEMS Yeast Res.* 11 (5): 408–415.

70 Olson, D.G. (2012). Enhanced cellulase expression in *Clostridium thermocellum. Appl Environ Microbiol.* 78 (18): 6739–6749.

71 Zhang, Y. (2020). Engineering *Aspergillus niger* for lignin degradation. *Enzyme Microb Technol.* 135: 109495.

72 Yadav, M. (2021). Ligninolytic enzyme expression in *Streptomyces coelicolor. J Environ Chem Eng.* 9 (4): 105384.

73 Nouri, H. (2022). Expression of *bglB* in *Zymomonas mobilis* for cellobiose conversion. *Renew Energy.* 184: 936–943.

74 Patel, A.K. (2020). Engineering *Rhizopus oryzae* for agro-waste bioconversion. *Waste Biomass Valor.* 11: 4739–4748.

75 Liu, Y. (2021). Enhanced lignocellulose degradation in MSW by engineered *Clostridium thermocellum. Renew Energy.* 164: 872–880.

76 Das, D. (2022). Biogas production using modified methanogens: current status and future perspectives. *Bioresour Technol.* 344: 126227.

77 Yu, J. (2020). Engineering *E. coli* for PHA production from organic MSW. *Bioresour Technol.* 314: 123734.

78 Singh, R. (2019). Laccase-mediated degradation of dyes by engineered *Bacillus subtilis. J Hazard Mater.* 375: 246–255.

79 Ward, P.G. (2021). Biodegradation of plastic residues using *Pseudomonas putida. Appl Microbiol Biotechnol.* 105 (3): 1051–1062.

80 Li, Z. (2020). Enhanced ethanol production in *S. cerevisiae* from food waste. *Renew Energy.* 148: 999–1006.

81 Olson, D.G. (2018). Genetically optimized *Clostridium thermocellum* for MSW biomass degradation. *Appl Environ Microbiol.* 84 (7): e02612-17.

82 Kumar, R. (2021). PET polymer degradation using *Aspergillus niger* expressing cutinase. *J Environ Chem Eng.* 9 (4): 105381.

83 Dietrich, K. (2019). Sustainable PHB production from MSW-derived VFAs using C. necator. *Biotechnol Bioeng.* 116 (5): 1183–1194.

84 Kook, M. (2022). Engineering *L. plantarum* for lactic acid fermentation of food waste. *Microb Cell Fact.* 21: 98.

85 Zhang, Y. (2018). Conversion of starch waste to organic acids using *R. oryzae. Waste Manag.* 72: 98–104.

86 Chauhan, A. (2021). Catabolism of aromatic MSW residues by engineered *Streptomyces lividans. Environ Sci Pollut Res.* 28 (29): 38781–38791.

87 Nies, D.H. (2016). Microbial heavy-metal resistance: bacteria with genetically engineered metal resistance genes. *J Indus Microbiol Biotechnol.* 43 (9): 1263–1278.

88 Sharma, B. (2021). Genetic enhancement of *Pseudomonas aeruginosa* for copper recovery from PCBs. *J Hazard Mater.* 413: 125413.

89 Juhas, M. (2014). Genome-based understanding of metal resistance in *R. metallidurans. FEMS Microbiol Rev.* 38 (4): 751–764.

90 Kumari, D. (2020). Genetically modified *E. coli* for precious metal recovery from e-waste. *Waste Manag.* 102: 618–626.

91 Li, Y. (2019). Citric acid-mediated bioleaching of metals from e-waste using *A. niger. Waste Biomass Valor.* 10 (11): 3153–3164.

92 Middleton, S.S. (2022). *S. oneidensis* engineered for reductive metal recovery. *Appl Environ Microbiol.* 88 (3): e02345-21.

93 Wu, Y. (2020). Biosorption of metals by engineered *S. cerevisiae. Bioresour Technol.* 319: 124204.

94 Dietrich, K. (2018). Bioplastic production from e-waste using *C. necator. J Environ Chem Eng.* 6 (5): 6246–6254.

95 Park, J. (2020). Engineered *Corynebacterium glutamicum* for rare earth recovery. *Biotechnol Bioeng.* 117 (7): 2139–2149.

96 Bose, R. and Jayant, M. (2021). *Yarrowia lipolytica* as a host for e-waste metal bioleaching. *Chemosphere.* 276: 130238.

97 Meruvu, H. (2021). Genetically modified microorganisms and their applications in environmental cleanup. *Int J Pharm Biol Sci.* 11 (3): 25–30.

98 Arshadi, M. and Yaghmaei, S. (2020). Bioremediation of waste using genetically engineered microorganisms. *Chemosphere*. 251: 126365.

99 Luo, G., Zhang, D., Li, Z., and Angenent, L.T. (2020). Genetically modified microorganisms for environmental biotechnology. *Biotechnol Adv*. 40: 107502.

100 Tsyganov, K., Donoghue, M.T., and Alexeyenko, A. (2020). Gene set and pathway enrichment analysis in genomics of genetically engineered microbes. *Brief Bioinform*. 21 (4): 1249–1263.

101 Sharma, B. (2020). Genetically engineered microbes: a tool for bioremediation. *J Environ Biotechnol Res*. 9 (1): 18–26.

102 Sharma, B. (2020). Environmental implications of genetically engineered microorganisms in waste treatment. *Biotech Front*. 4 (2): 33–42.

103 Brown, D., Leach, A., Fairchild, T. (2020). Designing safe and effective genetically engineered microbes for bioremediation. *Front Bioeng Biotechnol*. 8: 885.

104 Ayoug, M., Basu, S., and Raj, A. (2020). Engineered microbial systems for environmental cleanup: current advances and future directions. *Bioresour Technol Rep*. 12: 100580.

105 Juhas, M., Eberl, L., and Church, G.M. (2014). Essential genes as antimicrobial targets and tools for synthetic biology. *Nat Rev Microbiol*. 10 (10): 779–790.

106 Kumar, M., Bhatia, S.K., Kim, J. (2021). Genetically engineered microbes for sustainable environmental remediation. *Sci Total Environ*. 796: 148926.

107 Wang, G., Huang, J., Li, T. (2022). Environmental applications of genetically engineered microorganisms: progress and perspectives. *Crit Rev Biotechnol*. 42 (4): 512–534.

108 El-Morsi, T.M., Daoud, J.I., and Hassanein, A.M. (2019). Biodegradation of environmental pollutants using genetically engineered bacteria. *J Environ Sci Health A Tox Hazard Subst Environ Eng*. 54 (13): 1183–1193.

109 Middleton, J., Eley, C., and Thomas, J. (2021). Applications of synthetic biology in bioremediation: a review of recent developments. *Environ Sci Process Impacts*. 23 (6): 729–745.

110 Wang, Y. (2021). High-performance microbial systems for biodegradation. *Environ Sci Technol*. 55 (12): 8142–8155.

111 Ahmed, M. (2020). Environmental sustainability through synthetic biology. *Curr Opin Biotechnol*. 66: 99–107.

112 Heinemann, J.A. (2013). Safety evaluation of genetically modified microorganisms. *Environ Biosaf Res*. 12 (1): 17–23.

113 OECD. (2019). *Safety Assessment of Genetically Engineered Microorganisms*. OECD Publishing.

114 Zhao, Y. (2020). Performance of engineered microbes in open environments. *Front Bioeng Biotechnol*. 8: 892.

115 Marris, C. (2015). Public views on synthetic biology and GMOs. *Nat Biotechnol*. 33 (9): 899–900.

116 Convention on Biological Diversity. (2000). *The Cartagena Protocol on Biosafety*. United Nations.

117 El Karoui, M. (2019). Building predictive models for microbial engineering. *Nat Rev Genet*. 20 (8): 551–564.

Part IV

Guidelines, Policies, and Laws

14

Waste Management Rules and Regulations

14.1 Introduction

Effective waste management is an essential component of sustainable development, public health preservation, and environmental protection. As the volume and complexity of waste increase globally due to rapid urbanization, industrialization, and population growth, the role of legal and regulatory frameworks has become increasingly significant. These frameworks are not merely administrative tools; they serve as the foundation upon which comprehensive waste governance systems are built, guiding the collection, transportation, treatment, and disposal of various types of waste.

Unmanaged or poorly managed waste poses serious threats to the environment and public health. Open dumping, unregulated incineration, and inadequate treatment of hazardous materials can lead to soil and groundwater contamination, air pollution, and the spread of infectious diseases. In urban centers, the proliferation of solid waste contributes to clogged drainage systems, urban flooding, and degradation of living conditions. Beyond environmental and health implications, improper waste handling affects economic growth by increasing public expenditure on healthcare and reducing the aesthetic and commercial value of land. It also disproportionately impacts vulnerable communities, exacerbating social inequalities and reducing overall quality of life.

The establishment of legal frameworks for waste management stems from the recognition of these multifaceted risks. These laws provide a formal structure that defines roles and responsibilities among government bodies, private sectors, and citizens. They also ensure the enforcement of environmental standards and introduce penalties for violations. A well-developed legal framework incorporates principles such as the "polluter pays" principle, extended producer responsibility (EPR), and the waste hierarchy (reduce, reuse, recycle), all of which aim to reduce waste generation at the source and promote circular economy practices.

Biological Processes and Techniques in Zero Waste Management, First Edition. Garima Awasthi and Mohit Nigam. Published 2026 by Wiley-VCH GmbH.

Globally, the evolution of waste management legislation has been shaped by environmental movements, technological advancements, and international agreements. In developed nations, the initial impetus for waste regulation came during the mid-20th century when industrial waste and increasing urban refuse began to pose visible hazards. For instance, the United States introduced the Resource Conservation and Recovery Act (RCRA) in 1976, which provided comprehensive guidelines for the management of both hazardous and nonhazardous waste. In the European Union (EU), the Waste Framework Directive laid the groundwork for integrated waste strategies across member states, emphasizing waste prevention, reuse, and environmentally sound disposal.

The global environmental agenda, influenced by summits such as the United Nations Conference on Environment and Development (1992) and agreements like the Basel Convention (1989), further pushed countries to align their domestic regulations with international norms. These milestones have gradually transformed waste management from a municipal concern into a globally monitored issue with transboundary implications [1].

In the Indian context, the evolution of waste regulation has followed a trajectory marked by growing environmental awareness and judicial intervention. Traditionally, waste disposal in India was governed by local municipal acts with limited scope and enforcement capabilities. However, the rising pressure from urban expansion, industrial growth, and public health crises prompted the central government to enact more comprehensive legislation. A major turning point came with the Environment (Protection) Act of 1986, which empowered the central government to take measures for environmental conservation, including waste regulation.

Subsequent rules tailored to specific waste streams were introduced, such as the Biomedical Waste Management Rules (1998), Hazardous Waste Management Rules (2008), E-Waste Management Rules (2011), and Solid Waste Management Rules (2016). These rules delineated the responsibilities of waste generators, local authorities, and treatment facilities, while also integrating principles like EPR and segregation at source.

The judiciary has also played a pivotal role in strengthening waste governance in India. Public interest litigations (PILs) filed in the Supreme Court and National Green Tribunal (NGT) have resulted in landmark judgments mandating stricter compliance and proactive governance. For instance, the Supreme Court's directives in the *Municipal Solid Waste Management* case led to the formulation of detailed waste handling protocols for urban local bodies (ULBs).

Despite the presence of a robust legal structure, challenges remain in implementation, monitoring, and public participation. Many local authorities face constraints related to technical capacity, financial resources, and institutional coordination. Informal sectors, which play a significant role in waste recycling in India, are often excluded from formal waste management systems. Addressing these gaps requires a dynamic legal framework that is responsive to emerging waste streams such as plastic waste, construction and demolition (C&D) debris, and hazardous industrial by-products.

In conclusion, legal frameworks are indispensable for ensuring that waste management practices align with broader environmental, health, and socio-economic goals. Their evolution, both globally and within India, reflects a growing recognition of the critical role that law and policy play in transitioning toward a more sustainable and equitable waste management system. Continued reforms, stakeholder engagement, and enforcement mechanisms will be essential in addressing the current and future challenges posed by waste in a rapidly changing world [2].

14.2 Classification of Waste and Regulatory Needs

Waste management is a complex task due to the diverse nature of waste generated by households, industries, healthcare facilities, and commercial establishments. To address this diversity effectively, waste is categorized into distinct types, each requiring specific regulatory, treatment, and disposal approaches. This classification ensures the safe, efficient, and environmentally sound handling of waste, minimizing its impact on public health and ecosystems. The following is a detailed explanation of the major waste categories, their unique characteristics, handling requirements, and the rationale for category-specific regulations (Figure 14.1) [3, 4].

14.2.1 Municipal Solid Waste

Definition and Sources

Municipal Solid Waste (MSW) refers to the everyday waste generated by households, commercial establishments, institutions, and small-scale industries. It includes biodegradable organic matter (food and garden waste), recyclable materials (paper, plastics, and metals), and inert waste (dust, silt, and construction debris).

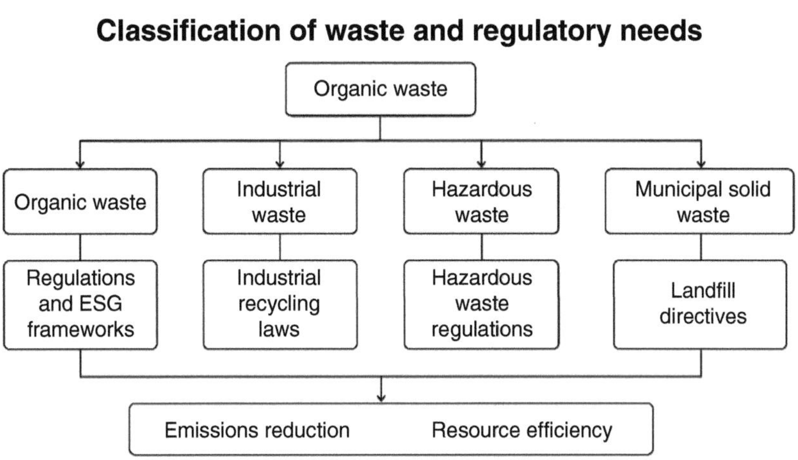

Figure 14.1 Flowchart of classification of waste and regulatory needs.

Handling and Disposal Needs

MSW requires systematic segregation at the source into biodegradable, recyclable, and nonrecyclable fractions. Composting and biomethanation are suitable for organic waste, while recyclable materials should be diverted to recycling units. Residual waste is directed to engineered sanitary landfills.

Regulatory Justification

The heterogeneous nature and large volume of MSW demand a well-regulated framework to prevent open dumping, reduce environmental contamination, and promote resource recovery. India's Solid Waste Management Rules, 2016 emphasize source segregation, decentralized processing, and involvement of waste pickers.

14.2.2 Hazardous Waste

Definition and Sources

Hazardous waste comprises materials that are toxic, corrosive, reactive, or flammable, posing a threat to human health and the environment. These wastes primarily originate from chemical manufacturing, pharmaceutical production, metallurgy, and petrochemical industries.

Handling and Disposal Needs

Hazardous waste requires meticulous storage, labeling, transportation, and treatment. It is often treated using methods like incineration, chemical neutralization, or stabilization before disposal in specially designed secured landfills to prevent leaching of toxins.

Regulatory Justification

Due to their potential for long-term environmental damage and occupational hazards, hazardous wastes are governed under stringent laws. The Hazardous and Other Wastes (Management and Transboundary Movement) Rules, 2016 regulate generation, import–export, treatment, and disposal, in compliance with international conventions like the Basel Convention.

14.2.3 Biomedical Waste

Definition and Sources

Biomedical waste includes any waste generated during diagnosis, treatment, or immunization of humans or animals. This includes sharps (needles, blades), soiled bandages, blood-soaked materials, discarded medicines, and microbiological waste from laboratories [5].

Handling and Disposal Needs

Biomedical waste must be segregated into color-coded containers at the point of generation. It is treated through autoclaving, incineration, or deep burial, depending on the type. Sharps are disinfected and shredded to prevent reuse.

Regulatory Justification

Improper handling of biomedical waste can lead to the transmission of infections such as HIV, Hepatitis B and C, and other communicable diseases. The Biomedical

Waste Management Rules, 2016 were enacted to ensure that such waste is handled safely, with clear responsibilities assigned to healthcare facilities and common treatment facilities.

14.2.4 E-waste

Definition and Sources
Electronic waste, or E-waste, consists of discarded electronic devices and components such as computers, mobile phones, televisions, refrigerators, and batteries. These items often contain valuable metals alongside toxic substances like lead, mercury, and cadmium.

Handling and Disposal Needs
E-waste needs to be collected separately and dismantled in authorized recycling facilities. Valuable components are recovered using safe processes, and hazardous elements are disposed of in compliance with environmental norms.

Regulatory Justification
The rapid growth of the electronics industry and short product lifespans have made E-waste one of the fastest-growing waste streams. The E-Waste (Management) Rules, 2016 emphasize EPR, mandating producers to take responsibility for collection and recycling.

14.2.5 Plastic Waste

Definition and Sources
Plastic waste encompasses discarded plastic materials such as bags, bottles, packaging, containers, and single-use products. It is generated across domestic, commercial, and industrial sectors.

Handling and Disposal Needs
Plastics should be segregated and sent for recycling or co-processing in cement kilns or waste-to-energy (WTE) plants. Thin and single-use plastics often require phase-out or bans due to their nonrecyclable nature.

Regulatory Justification
Plastic pollution, especially microplastics, has severe environmental implications, including harm to marine life and food chain contamination. The Plastic Waste Management Rules, 2016 (amended in 2021) promote segregation, recycling, reuse, and a ban on certain single-use plastics, while mandating EPR compliance from producers and brand owners [6].

14.2.6 C&D Waste

Definition and Sources
C&D waste includes debris from construction, renovation, and demolition activities such as concrete, bricks, steel, wood, and plumbing materials. It is a bulky waste stream that significantly contributes to urban waste.

Handling and Disposal Needs
C&D waste needs to be collected separately, and materials like concrete and bricks can be crushed and reused as road base or construction aggregates. Hazardous components (e.g., asbestos) require special handling.

Regulatory Justification
The large volume of C&D waste, if unmanaged, can cause urban congestion, illegal dumping, and environmental degradation. The Construction and Demolition Waste Management Rules, 2016 mandate segregation at source, establishment of recycling facilities, and utilization of recycled materials in construction projects.

14.3 National Waste Management Rules in India (2016 Onward)

In 2016, the Government of India introduced a series of comprehensive waste management rules aimed at strengthening environmental governance and promoting sustainable waste practices across the country. These rules, notified under the Environment (Protection) Act, 1986, address the specific handling requirements of various waste streams and emphasize the principles of segregation at source, EPR, and scientific processing and disposal. The 2016 rules represented a significant overhaul of earlier regulations, aligning them with modern challenges such as urban expansion, industrial growth, and the need for circular economy practices [7].

14.3.1 Overview of Major 2016 Rules and Their Amendments

14.3.1.1 Solid Waste Management Rules, 2016

These rules replaced the earlier Municipal Solid Waste (Management and Handling) Rules, 2000. Key features include:

- Mandatory segregation of waste at source into biodegradable, nonbiodegradable, and domestic hazardous waste.
- Promotion of decentralized waste processing methods such as composting and biomethanation.
- Inclusion of bulk waste generators, gated communities, and institutions in waste management responsibilities.
- Emphasis on integrating informal waste pickers into the formal system.

Amendments
Later revisions focused on waste processing targets, clarity on compost quality, and strengthening the responsibilities of local bodies.

14.3.1.2 Hazardous and Other Wastes (Management and Transboundary Movement) Rules, 2016

These rules regulate the safe handling, storage, treatment, and disposal of hazardous wastes and their import/export.

- Clear classification of hazardous waste based on physical and chemical characteristics.
- Requirement for generators to obtain authorization from the State Pollution Control Boards (SPCBs).
- Provisions for co-processing of hazardous waste in cement kilns and other industries.

Amendments

Updates have been made to streamline import/export documentation and expand the list of hazardous waste types.

14.3.1.3 Biomedical Waste Management Rules, 2016

These rules aim to ensure safe handling and disposal of biomedical waste from healthcare facilities [8].

- Segregation of waste into color-coded bins at the point of generation.
- Mandatory pretreatment of certain waste types like microbiological waste and sharps.
- Use of barcoding and GPS tracking for waste movement.
- Authorization required for all healthcare facilities, including small clinics and veterinary institutions.

Amendments

Focused on stricter timelines for compliance, integration of tracking technologies, and improved monitoring by State authorities.

14.3.1.4 Plastic Waste Management Rules, 2016

These rules replaced the 2011 version and aimed to tackle the growing problem of plastic pollution.

- Mandatory registration of plastic producers, importers, and brand owners.
- Implementation of EPR, requiring producers to collect and manage post-consumer plastic.
- Ban on plastic bags below 50 microns in thickness.

Amendments (especially 2021 and 2022)

- Phased ban on identified single-use plastic items.
- Introduction of national EPR guidelines with annual recycling targets and penalties for noncompliance [9].

14.3.1.5 E-Waste (Management) Rules, 2016

These rules address the environmentally sound disposal of electronic and electrical waste.

- Establishment of Producer Responsibility Organizations (PROs) to collect e-waste on behalf of producers.

- Compulsory registration of e-waste recyclers and dismantlers.
- Emphasis on environmentally sound recycling practices and safe recovery of valuable materials.

Amendments (notably in 2022)

- Shift towards a centralized digital portal for EPR registration and reporting.
- Clear annual recycling targets and penalties for shortfalls.

14.3.1.6 Construction and Demolition Waste Management Rules, 2016

These rules cover the collection, transportation, processing, and reuse of waste generated during C&D activities.

- Requirement for segregation of C&D waste at source into concrete, metal, wood, etc.
- Mandate for reuse of recycled materials in construction projects.
- ULBs are to set up collection centers and recycling plants [10].

Amendments

Have encouraged greater private sector involvement and set utilization targets for recycled materials.

14.3.2 Roles of Stakeholders

1. **Urban Local Bodies (ULBs)**
   - Primary responsibility for collection, transportation, and processing of municipal waste.
   - Enforcement of segregation, user fees, and setting up infrastructure for composting and recycling.
   - Monitoring bulk generators and facilitating public awareness campaigns.
2. **State Pollution Control Boards (SPCBs)**
   - Granting authorization to waste generators, processors, and treatment facilities.
   - Periodic monitoring and inspection of compliance by different stakeholders.
   - Submission of annual performance reports to the Central Pollution Control Board (CPCB).
3. **Central Pollution Control Board (CPCB)**
   - Lays down national-level guidelines and technical standards.
   - Maintains databases and tracks compliance across states.
   - Provides training and technical support to SPCBs and other stakeholders.
4. **Producers, Recyclers, and Waste Generators**
   - Producers are mandated to fulfill EPR obligations under Plastic waste and E-waste rules.
   - Recyclers and dismantlers must obtain authorization and follow environmentally sound processing methods.
   - Waste generators, including households and institutions, are responsible for segregating waste at source and avoiding littering [11].

14.3.3 Compliance Mechanisms and Enforcement Agencies

The enforcement of waste management rules is carried out through a multitiered system:

- **Authorization Mechanism:** SPCBs issue authorizations to waste handlers, requiring compliance with environmental norms.
- **Monitoring and Reporting:** Regular inspections, audits, and submission of annual compliance reports are mandated.
- **Penalties and Legal Action:** Noncompliance attracts penalties under the Environment (Protection) Act, 1986, including fines and closure notices.
- **Technological Tools:** Use of barcoding, GPS tracking, and online portals (especially in biomedical and e-waste management) enhances traceability and transparency.
- **Capacity Building:** CPCB and SPCBs conduct training programs and workshops for municipal authorities and waste operators.

The suite of waste management rules introduced in 2016 has laid a strong legal foundation for structured, efficient, and environmentally responsible waste governance in India. These rules, supported by stakeholder-specific roles and robust compliance mechanisms, aim to transition the country from a linear "use and discard" model to a circular, resource-efficient economy. Continued enforcement, technological integration, and public participation will be crucial to realizing the full potential of these frameworks [12].

14.4 International Regulatory Landscape and Agreements

The global nature of waste generation and environmental pollution necessitates international cooperation, harmonized policies, and legally binding agreements. Several international conventions, national laws, and regional directives have emerged to address the multifaceted challenges associated with waste management—particularly hazardous and persistent wastes. These legal instruments not only facilitate the safe handling, transportation, and disposal of waste but also promote accountability, sustainability, and innovation in waste governance. Following are the most significant international frameworks and policies shaping the global waste management landscape.

14.4.1 Basel Convention on the Control of Transboundary Movements of Hazardous Wastes and Their Disposal

Adopted in 1989 and coming into force in 1992, the Basel Convention is a cornerstone of global environmental governance in the context of hazardous waste. Its primary objective is to reduce the movement of hazardous wastes across borders, especially from developed to developing countries, and to ensure that such waste is managed and disposed of in an environmentally sound manner [13].

Key provisions include:

- Prior Informed Consent (PIC) mechanism for cross-border waste transfer.
- Strict documentation and tracking of waste shipments.
- Encouragement of waste minimization at the source.
- Prohibition of export of hazardous waste to countries lacking the capacity to manage it safely.

India ratified the Basel Convention in 1992, and its national hazardous waste regulations reflect many of its principles.

14.4.2 Stockholm Convention on Persistent Organic Pollutants (POPs)

The Stockholm Convention, signed in 2001 and enforced in 2004, targets Persistent Organic Pollutants—chemical substances that persist in the environment, bioaccumulate in living organisms, and pose serious threats to human health and ecosystems. Common examples include polychlorinated biphenyls (PCBs), dioxins, and certain pesticides.

Major commitments under the Convention:

- Elimination or restriction of production and use of listed POPs.
- Environmentally sound management and disposal of POP stockpiles and wastes.
- Promotion of safer chemical alternatives and substitution of hazardous substances.

The convention has been instrumental in guiding nations to phase out highly toxic chemicals and establish safe disposal methods for contaminated waste.

14.4.3 EU Waste Framework Directive

The EU Waste Framework Directive (Directive 2008/98/EC) forms the legal basis for waste management across the EU. It introduces the concept of the waste hierarchy, which prioritizes:

1. Waste prevention,
2. Reuse,
3. Recycling,
4. Recovery (including energy recovery), and
5. Disposal as the last resort.

Other important aspects include:

- The polluter pays principle, requiring waste producers to bear the cost of management.
- Life cycle thinking, encouraging sustainable design and material use.

- Mandatory waste planning and reporting by member states.
- Clear definitions and classifications of waste to harmonize legislation across borders [14].

The Directive has catalyzed innovation in recycling technologies and advanced the circular economy in Europe.

14.4.4 United States Resource Conservation and Recovery Act

RCRA, enacted in 1976, is a key legislative framework governing waste management in the United States. RCRA addresses both hazardous and nonhazardous solid waste and promotes the "cradle-to-grave" approach—ensuring full accountability from the point of waste generation to its final disposal.

Key components include:

- Regulations for waste identification, classification, and listing.
- Requirements for waste treatment, storage, and disposal facilities.
- Monitoring and reporting obligations for hazardous waste generators and transporters.
- Encouragement of source reduction and resource recovery programs.

RCRA has significantly contributed to minimizing illegal dumping and enhancing transparency in waste tracking within the United States

14.4.5 EPR in the Global Context

EPR is a policy approach where producers are made financially and/or physically responsible for the entire life cycle of their products, particularly post-consumer disposal. EPR is widely recognized as a key driver for sustainable product design and effective waste collection systems.

Global adoption examples:

- Germany introduced the Packaging Ordinance in the 1990s, one of the earliest EPR laws, which mandated producers to take back and recycle packaging waste.
- Japan implemented the Home Appliance Recycling Law, requiring manufacturers to recover and recycle specific electronic goods.
- Canada has multiple provincial EPR programs for products ranging from electronics to household hazardous waste [15].

EPR schemes have proven effective in reducing the burden on municipal systems, improving recycling rates, and encouraging eco-design. Many developing countries, including India, are increasingly incorporating EPR into their legal framework for plastics, e-waste, and batteries.

14.5 Implementation Challenges and Gaps

Despite the establishment of comprehensive legal frameworks and national waste management rules in India and across the globe, the effective implementation of these policies faces several challenges. These obstacles span administrative inefficiencies, infrastructural limitations, socio-economic dynamics, and weak enforcement mechanisms. Addressing these gaps is crucial for achieving environmentally sound and inclusive waste governance [16].

14.5.1 Administrative, Infrastructural, and Legal Barriers

One of the most pressing challenges in implementing waste management rules is the lack of administrative coordination between different government departments and agencies. ULBs, which are primarily responsible for managing MSW, often suffer from limited technical expertise, workforce shortages, and inadequate funding. In many municipalities, solid waste collection and processing infrastructure such as composting units, material recovery facilities (MRFs), and sanitary landfills are either absent or insufficient.

On the legal front, ambiguities in overlapping responsibilities between central, state, and local authorities often lead to regulatory fragmentation and weak enforcement. Additionally, frequent amendments in rules without corresponding capacity-building measures for ground-level officials result in confusion and poor compliance.

14.5.2 Informal Sector Integration

The informal waste sector plays a vital role in India's recycling economy, particularly in the collection, sorting, and resale of recyclable materials such as plastics, paper, and metals. However, waste pickers and scrap dealers often operate outside the formal regulatory framework. Their exclusion leads to social insecurity, health risks, and a missed opportunity to strengthen decentralized waste processing systems.

Efforts to integrate the informal sector into formal waste management plans have been slow and inconsistent. Challenges include lack of identification documents, resistance from organized waste handlers, and limited access to protective gear or financial support. Proper integration would not only enhance recycling efficiency but also promote social inclusion and livelihoods for marginalized communities.

14.5.3 Issues of Compliance in Rural and Peri-urban Areas

Waste management policies have largely been urban-centric, leaving rural and peri-urban areas underserved. In many villages and small towns, systematic collection and segregation of waste are not practiced. Open dumping, burning of waste, and unregulated use of plastics are common, posing significant environmental and public health hazards.

Moreover, awareness about waste rules and environmental responsibilities remains low in these areas. The absence of tailored policies, infrastructure, and trained manpower exacerbates the problem, resulting in poor implementation of national rules in nonurban regions.

14.5.4 Monitoring, Auditing, and Reporting Lapses

Effective monitoring and reporting are essential to ensure that stakeholders comply with waste management regulations. However, these functions often suffer from systemic gaps. Many ULBs and SPCBs do not have robust systems to track waste generation, segregation rates, or recycling performance. Annual reports required under different waste rules are either incomplete or not submitted on time.

Technological solutions such as GPS tracking of waste vehicles, barcoding of biomedical waste, and online compliance portals have been introduced, but their adoption is uneven across states. Moreover, there is limited auditing of private sector compliance under EPR, especially in sectors like plastic and e-waste, allowing many producers to evade their obligations [17].

14.5.5 Emerging Trends and Policy Innovations

In response to the growing complexity of waste generation and environmental degradation, waste management systems are evolving through the adoption of innovative policies and technologies. One of the most transformative concepts gaining momentum is the circular economy model, which aims to shift from a linear "take-make-dispose" approach to one that emphasizes reuse, recycling, and resource recovery. This model promotes sustainable product design, extended product life cycles, and closed-loop supply chains, thereby minimizing waste generation at the source. Complementing this shift are smart waste tracking systems, which use digital tools such as barcodes, RFID tags, and GPS to monitor the movement of waste from generation to disposal. These systems enhance transparency, improve regulatory compliance, and enable real-time data collection for better decision-making.

Governments are also introducing incentive-based regulatory mechanisms to promote responsible waste behavior. The "polluter pays" principle ensures that entities generating waste bear the cost of its management, encouraging waste reduction at the source. Meanwhile, deposit-refund schemes offer consumers a financial incentive to return products or packaging materials for proper recycling, as seen in bottle-return programs in several countries. Another area witnessing rapid advancement is the use of emerging technologies like Artificial Intelligence (AI), Internet of Things (IoT), and blockchain in waste regulation. AI-powered tools can predict waste generation patterns and optimize collection routes, while IoT-enabled bins and sensors monitor waste levels and improve operational efficiency. Blockchain technology, with its decentralized and tamper-proof nature, is being explored to create transparent records of waste transactions, particularly in recycling and EPR compliance [18].

Additionally, the intersection of waste management and climate policy is becoming increasingly relevant, especially through carbon credit systems and WTE frameworks. Facilities that convert waste into energy, through incineration, anaerobic digestion, or gasification, can generate renewable energy while also reducing landfill use. When designed efficiently and with environmental safeguards, these systems can contribute to carbon offset programs, enabling municipalities and industries to earn credits for reducing greenhouse gas emissions. Overall, these emerging trends and innovations signify a paradigm shift in how waste is perceived, not as a liability, but as a valuable resource, thereby, aligning environmental goals with economic and technological progress.

14.5.6 Case Studies and Comparative Analysis

Understanding how different regions and cities approach waste management provides valuable insights into what strategies work best under varying socio-economic and regulatory conditions. Comparative studies and real-world success stories reveal critical lessons in effective policy design, citizen participation, technological integration, and the role of partnerships in driving sustainability. Following are key case studies and comparisons that highlight the global and local dimensions of waste governance.

14.5.6.1 India Versus EU: Waste Segregation and Recycling Benchmarks

The EU is recognized globally for its advanced waste management systems, driven by strong legal mandates and high levels of public compliance. Under the EU's Waste Framework Directive (2008/98/EC), member states are required to follow the waste hierarchy: prevention, reuse, recycling, recovery, and disposal. Most EU countries have implemented mandatory source segregation of waste into multiple streams, organic, paper, plastics, glass, and hazardous waste. As a result, recycling rates are significantly high; for example, Germany has achieved a municipal waste recycling rate of over 65%, with widespread citizen participation and robust EPR systems.

In contrast, India has made considerable progress in recent years with the implementation of the Solid Waste Management Rules, 2016, which mandate segregation of waste at source into biodegradable, nonbiodegradable, and domestic hazardous categories. However, segregation compliance remains inconsistent, particularly in tier-2 and rural areas. While urban centers like Indore and Pune have achieved over 80% segregation levels through strong civic engagement and infrastructure development, the national average remains below 30%. Moreover, recycling in India is largely driven by the informal sector, which lacks formal recognition and support. The EU's model demonstrates how strict enforcement, technological support, and civic education can significantly boost segregation and recycling, offering a pathway for India to emulate [19].

14.5.6.2 E-waste Handling: India Versus Japan

Electronic waste, or e-waste, presents one of the fastest-growing environmental challenges globally due to the rapid pace of technological advancement and consumerism. Japan has been a leader in e-waste management, employing a highly structured system under the Home Appliance Recycling Law and other related legislation.

Producers and consumers share responsibility, and specific electronic items, such as televisions, air conditioners, and washing machines, must be returned to designated collection centers. The government enforces stringent recycling standards and promotes product designs that are easier to dismantle and recycle. Citizens are well-informed about their responsibilities, and the entire process is tightly monitored.

In India, the E-Waste (Management) Rules, 2016 established the framework for formal handling of e-waste and introduced the concept of EPR. While these rules mark a progressive step, implementation challenges remain. Many producers fail to meet collection targets, informal recycling continues to dominate the sector, and public awareness of safe disposal methods is still low. Unlike Japan's well-regulated system, India struggles with traceability, unauthorized recycling units, and improper disposal, leading to environmental and health hazards. Strengthening enforcement, supporting formal recyclers, and increasing public education could help bridge this gap [20].

14.5.6.3 Role of Public–Private Partnerships in Improving Rule Compliance

Public–Private Partnerships (PPPs) have emerged as an effective model to overcome financial and operational barriers in waste management. By leveraging private sector efficiency and public sector oversight, PPPs can deliver improved service quality, innovation, and infrastructure development.

A notable example is the WTE plant at Okhla, Delhi, which operates under a PPP framework. Despite initial controversies regarding emissions, the project has showcased how private investment can enable large-scale waste processing in metropolitan areas. Similarly, Pune's SWaCH Cooperative, a collaboration between the Pune Municipal Corporation and waste picker cooperatives, represents a unique public–community–private model. Here, informal workers are formally contracted to collect segregated waste directly from households, ensuring employment, improving segregation rates, and reducing landfill dependence.

In the EU, PPPs are standard practice, particularly in recycling and composting infrastructure development. For instance, several municipalities in Italy and France have outsourced waste collection and processing to private contractors under strict environmental and service-level agreements. These partnerships are often tied to performance metrics, which incentivize efficiency and accountability.

In India, expanding PPPs in secondary waste processing, biomethanation, composting units, and collection logistics—especially in smaller towns—can help improve compliance with waste management rules and reduce dependency on government funding alone [21].

14.5.6.4 Successful Municipal Models: Indore and Pune

India offers several successful municipal case studies that demonstrate how localized efforts, citizen engagement, and administrative will can transform waste management systems.

Indore: The Cleanest City in India

Indore has consistently ranked first in the Swachh Survekshan survey for several years, largely due to its comprehensive waste management model. The city has implemented 100% door-to-door collection with source segregation, real-time GPS tracking of waste collection vehicles, and a well-structured MRF. The Indore Municipal Corporation also operates composting units and a biomethanation plant, ensuring that almost all organic waste is scientifically processed. Additionally, public engagement campaigns, strict fines for littering, and efficient grievance redressal systems have played a major role in transforming public behavior. Indore's success lies in its integration of technology, public participation, and strong governance [22].

Pune: A Model of Informal Sector Inclusion
Pune stands out for its inclusive waste management approach, particularly through the work of the SWaCH Cooperative, which empowers waste pickers by integrating them into the formal system. Waste pickers, trained and registered, collect segregated waste directly from households, improving both collection efficiency and recycling rates. The Pune Municipal Corporation supports this model by providing uniforms, ID cards, and basic health coverage, while allowing waste pickers to charge a nominal fee from residents. This system not only ensures better compliance with the Solid Waste Management Rules but also promotes social equity and livelihood security.

Both Indore and Pune highlight how localized innovation, strong institutional support, and stakeholder collaboration can overcome structural challenges and serve as replicable models for other cities [23].

14.6 Recommendations and Future Directions

Effective waste management requires a multidimensional approach that addresses policy enforcement, technological advancement, and behavioral change. While significant progress has been made in establishing legal frameworks and waste handling guidelines, the current system faces implementation gaps that demand actionable improvements. The following are key recommendations to strengthen waste governance in India and align it with international best practices, followed by a section on future directions that can drive long-term sustainability.

14.6.1 Strengthening Enforcement Mechanisms

Robust enforcement is essential to ensure compliance with waste management laws. Despite the presence of comprehensive rules, weak implementation often results in illegal dumping, poor segregation, and noncompliance with EPR. Strengthening the institutional capacity of regulatory bodies such as the CPCB and SPCBs is critical. This includes increasing staffing, improving inspection protocols, and leveraging digital tools for monitoring. Introduction of stricter penalties, periodic audits, and independent third-party assessments can further enhance regulatory compliance [24].

14.6.2 Capacity Building and Stakeholder Education

There is an urgent need to enhance the capabilities of local authorities, sanitation workers, and informal waste handlers. Training programs, technical workshops, and certification modules can help municipal officials better understand and implement waste management rules. Educational campaigns targeting producers, recyclers, healthcare facilities, and construction companies should focus on rule-specific responsibilities and environmentally sound practices. Additionally, integrating waste management topics into school and college curricula can foster long-term awareness and responsibility [25].

14.6.3 Regulatory Harmonization with Global Standards

To align with international norms, India must harmonize its waste regulations with conventions such as the Basel and Stockholm Conventions and adopt best practices from developed countries. This includes streamlining waste classification, enhancing reporting mechanisms, and enforcing EPR frameworks that mirror the standards followed in the EU or Japan. International collaboration and knowledge exchange can also help India build resilient systems capable of addressing complex waste streams like e-waste and hazardous industrial residues [26].

14.6.4 Promotion of Green Technologies and Innovations

Encouraging innovation in waste processing and recycling can significantly reduce environmental impacts and increase resource efficiency. The government should provide incentives—such as tax breaks, subsidies, and grants—for industries adopting green technologies, such as anaerobic digestion, bioremediation, plastic-to-fuel conversion, and AI-based waste sorting systems. Establishing waste innovation hubs, start-up incubators, and PPPs can stimulate R&D in sustainable waste solutions [27].

14.6.5 Citizen Participation and Behavioral Change Strategies

Public engagement is central to the success of any waste management system. Efforts should be made to foster behavioral change through sustained awareness campaigns, community outreach, and reward-based incentives. For example, households and institutions that practice effective waste segregation or composting can be recognized with rebates or certifications. Mobile apps, local workshops, and neighborhood competitions can encourage active participation and create a culture of environmental responsibility [28].

14.7 Future Directions

Looking ahead, the evolution of waste governance will rely on integrated and forward-thinking strategies that go beyond regulatory compliance. The following future directions are critical for building a resilient, inclusive, and circular waste economy:

- **Digital Transformation:** Adoption of advanced technologies such as AI, IoT, and blockchain will transform waste tracking, data management, and compliance reporting. Real-time monitoring systems can improve accountability and streamline logistics.

- **Circular Economy Integration:** Embedding circular economy principles into national and local policies will help minimize waste generation and maximize material recovery. This includes designing products for durability, repairability, and recyclability.
- **Climate-Smart Waste Management:** Waste strategies should be aligned with climate action plans by promoting WTE, reducing methane emissions from landfills, and integrating carbon credit systems.
- **Inclusive Policy Frameworks:** Future policies must recognize the role of informal waste workers and ensure their integration through legal recognition, social protection, and skill development.
- **Decentralized Waste Solutions:** Promoting localized, community-based waste management models—such as decentralized composting or biogas units—can reduce pressure on urban infrastructure and foster grassroots participation.

14.8 Conclusion

Effective regulation is the backbone of any comprehensive waste management strategy. Legal frameworks not only provide the structural foundation for waste segregation, handling, and disposal, but also define the responsibilities of various stakeholders—from policymakers and industries to municipalities and citizens. Over the years, the evolution of national and international waste management laws has played a critical role in addressing the environmental, health, and socio-economic challenges posed by unregulated waste.

However, the mere existence of rules is not sufficient. Ensuring sustainable and inclusive implementation requires a multipronged approach that includes strong enforcement, capacity building, and active public engagement. Integrating informal waste sectors, promoting behavioral change, and encouraging green technologies are essential for translating regulations into effective action. Equally important is the harmonization of domestic policies with global environmental agreements, which can elevate national practices to internationally recognized standards.

As the volume and complexity of waste streams continue to grow—particularly with the rise in e-waste, plastics, and hazardous residues—there is a pressing need for dynamic and adaptive policy frameworks. These must be responsive to emerging challenges and innovations, allowing for real-time adjustments and continuous improvement. Moving forward, the focus must shift from reactive waste disposal to proactive waste prevention and resource recovery. By fostering a collaborative, tech-enabled, and environmentally conscious regulatory ecosystem, India and other nations can pave the way for a more resilient and circular waste economy.

References

1 Ministry of Environment, Forest and Climate Change (2016). *Solid Waste Management Rules, 2016*. Gazette of India.

2 Ministry of Environment, Forest and Climate Change (2016). *Hazardous and Other Wastes (Management and Transboundary Movement) Rules, 2016*. Gazette of India.

3 Ministry of Environment, Forest and Climate Change (2016). *Bio-Medical Waste Management Rules, 2016*. Gazette of India.

4 Ministry of Environment, Forest and Climate Change (2016). *E-Waste (Management) Rules, 2016*. Gazette of India.

5 Ministry of Environment, Forest and Climate Change (2016). *Plastic Waste Management Rules, 2016*. Gazette of India.

6 Ministry of Environment, Forest and Climate Change (2016). *Construction and Demolition Waste Management Rules, 2016*. Gazette of India.

7 Central Pollution Control Board (2022). *Guidelines for Implementation of E-Waste Rules, 2022*. Delhi: CPCB.

8 Central Pollution Control Board (2023). *Annual Report on Implementation of Plastic Waste Management Rules*. New Delhi: CPCB.

9 European Commission (2008). *Directive 2008/98/EC on Waste (Waste Framework Directive)*. Official Journal of the European Union.

10 United Nations Environment Programme (1989). *Basel Convention on the Control of Transboundary Movements of Hazardous Wastes and Their Disposal*. Geneva: UNEP.

11 United Nations Environment Programme (2001). *Stockholm Convention on Persistent Organic Pollutants*. Geneva: UNEP.

12 United States Environmental Protection Agency (1976). *Resource Conservation and Recovery Act (RCRA)*. Washington DC: EPA.

13 Ministry of Housing and Urban Affairs (2023). *Swachh Bharat Mission – Urban*. Annual Progress Report.

14 Japan Ministry of Environment (2001). *Law for the Promotion of Effective Utilization of Resources*. Tokyo: MoE.

15 Ghosh, S.K. (2020). Waste Management Laws in India: A Commentary. New Delhi: LexisNexis.

16 Gupta, A. and Singh, R. (2022). Legal framework and challenges of e-waste management in India. *Environ Sci Pollut Res*. 29 (14): 20098–20105.

17 Sharma, B., Jain, P., and Garg, P. (2021). Emerging trends in solid waste management: a review. *Waste Manag*. 121: 172–184.

18 Kumar, A. and Agrawal, A. (2020). Current status and future perspectives of municipal solid waste management in India. *Renew Sustain Energy Rev*. 120: 109651.

19 United Nations Industrial Development Organization (2022). *Public-Private Partnerships for Waste Management*. Vienna: UNIDO.

20 Ellen MacArthur Foundation (2013). *Towards a Circular Economy: Economic and Business Rationale for an Accelerated Transition.* Cowes, UK: Ellen MacArthur Foundation.

21 Kaza, S., Yao, L., Bhada-Tata, P., and Van Woerden, F. (2018). *What a Waste 2.0: A Global Snapshot of Solid Waste Management to 2050.* Washington, DC: World Bank.

22 Paritosh, K., Yadav, M., Vivekanand, V. et al. (2016). Bio-medical waste management: a review. *J Clin Diagn Res.* 10 (6): LE01–LE04.

23 Borthakur, A. and Govind, M. (2018). E-waste management in India: a stakeholder's perspective. *Environ Dev Sustain.* 20: 2317–2330.

24 Singh, R., Tyagi, V., and Allen, T. (2023). Role of technology in smart waste management: a review. *J Environ Manage.* 336: 117633.

25 Central Pollution Control Board (2022). *Performance Evaluation of Solid Waste Management in Indian Cities.* Delhi: CPCB.

26 Ministry of Environment, Forest and Climate Change (2017). *Guidelines for Environmentally Sound Management of Construction and Demolition (C&D) Wastes.* New Delhi: Government of India.

27 Chaturvedi, B. and Sharma, S. (2022). Public awareness and behavioural change for sustainable waste management in India. *Indian J Environ Prot.* 42 (4): 345–350.

28 Indore Municipal Corporation (2023). *Smart Waste Management Case Study.* Indore Smart City Dashboard.

Index

Note: Page numbers in *italics* and **bold** refers to figures and tables respectively.